*Construction*
*Specifications Writing:*
*Principles and Procedures*

*Wiley Series of Practical Construction Guides*

# M. D. MORRIS, P.E., EDITOR

Jacob Feld
CONSTRUCTION FAILURE

William G. Rapp
CONSTRUCTION OF STRUCTURAL STEEL BUILDING
FRAMES

John Philip Cook
CONSTRUCTION SEALANTS AND ADHESIVES

Ben C. Gerwick, Jr.
CONSTRUCTION OF PRESTRESSED CONCRETE
STRUCTURES

S. Peter Volpe
CONSTRUCTION MANAGEMENT PRACTICE

Robert Crimmins, Reuben Samuels, and Bernard Monahan
CONSTRUCTION ROCK WORK GUIDE

B. Austin Barry
CONSTRUCTION MEASUREMENTS

D. A. Day
CONSTRUCTION EQUIPMENT GUIDE

Harold J. Rosen, P.E., FCSI
CONSTRUCTION SPECIFICATIONS WRITING:
PRINCIPLES AND PROCEDURES

Gordon Fletcher and Al Smoots
CONSTRUCTION GUIDE FOR SOILS AND FOUNDATIONS

# Construction
# Specifications Writing
## Principles and Procedures

*Harold J. Rosen, P.E., FCSI*

A Wiley-Interscience Publication

*JOHN WILEY & SONS, INC.* New York • *London* • *Sydney* • *Toronto*

Copyright © 1974, by John Wiley & Sons, Inc.

All rights reserved. Published simultaneously in Canada.

No part of this book may be reproduced by any means, nor transmitted, nor translated into a machine language without the written permission of the publisher.

**Library of Congress Cataloging in Publication Data:**

Rosen, Harold J
  Construction specifications writing.

  (Wiley series of practical construction guides)
  "A Wiley-Interscience publication."
  1. Specifications writing. 2. Building—Contracts
and specifications.  I. Title.
TH425.R59      692'.3      73-16138
ISBN 0-471-73551-5

Printed in the United States of America

10 9 8 7 6 5 4 3

# Series Preface

The construction industry in the United States and other advanced nations continues to grow at a phenomenal rate. In the United States alone construction in the near future will exceed ninety billion dollars a year. With the population explosion and continued demand for new building of all kinds, the need will be for more professional practitioners.

In the past, before science and technology seriously affected the concepts, approaches, methods, and financing of structures, most practitioners developed their know-how by direct experience in the field. Now that the construction industry has become more complex there is a clear need for a more professional approach to new tools for learning and practice.

This series is intended to provide the construction practitioner with up-to-date guides which cover theory, design, and practice to help him approach his problems with more confidence. These books should be useful to all people working in construction: engineers, architects, specification experts, materials and equipment manufacturers, project superintendents, and all who contribute to the construction or engineering firm's success.

Although these books will offer a fuller explanation of the practical problems which face the construction industry, they will also serve the professional educator and student.

M. D. MORRIS, P.E.

# *Preface*

Specifications writing in recent years has witnessed some dramatic changes as it is slowly evolving from an art to the semblance of a science. Prior to World War II when the number of materials available to the design professional was rather limited, specifications and the specifier were both held in rather low esteem. The process of specifying was usually performed by taking on older specifications as similar as possible to the project at hand and crossing out, filling in, and generally cutting and pasting together a patchwork of specification clauses to represent the new project specification.

With the advent of the materials explosion following World War II, specifications writing suddenly required more talented individuals to prepare specifications and more sophisticated techniques to handle the storage and retrieval of voluminous data. Both the individual and the techniques are now receiving the attention due them. Educational courses at the college level and programs aimed at updating the knowledge of the active specifier are multiplying to adequately prepare the individual for his more demanding role. This book is intended as an additional tool for educators to supplement the standard courses in specifications writing. Master specifications and computerization of specifications are discussed in this text as a means of strengthening the techniques required for processing and preparing specifications.

The proliferation of materials and the search for new design solutions have placed added emphasis on the need for developing a systematic approach to the evaluation and selection of materials. Accordingly, a chapter in this book deals with this subject in a comprehensive manner to guide the specifier through the selection process.

The advent of systems building has created a similar need for improvement in the area of performance specifications to allow manufacturers a freer hand in the evolution of products and systems to meet performance criteria. The most recent developments in this area are described herein.

The basic principles and guide lines set forth in this book enable the student or beginner to understand more readily the basic concept of specifications writing that I have gleaned and amassed through various means, including over thirty years of practice, my articles on specifications that have appeared in *Progressive Architecture* since 1956 in the column "Specifications Clinic," and standards developed by the Construction Specifications Institute and the American Institute of Architects. In addition, by outlining specific procedures, the student and practitioner are taken step by step through the stages necessary to prepare specifications.

This book crystallizes and codifies these various elements that are essential to a better comprehension of *construction specifications writing* and to a method of producing them by separating the information into two distinct aspects, *principles and procedures.*

HAROLD J. ROSEN

*Merrick, New York*
*September 1973*

# Contents

Introduction to Principles and Procedures
Specifications Writing

## PRINCIPLES

## BIDDING PROCEDURES

x    Contents

*Construction*
*Specifications Writing:*
*Principles and Procedures*

# Introduction to Principles and Procedures of Specifications Writing

A knowledge of specification writing principles and procedures is essential to the specifier in the architect's and engineer's office in the preparation of sound, enforceable specifications. Unless these skills are properly developed, an expert knowledge of materials, contracts, and construction procedures cannot be communicated successfully to the ultimate users of the finished specifications. The users, namely, contractors and materials suppliers, will also have a better understanding of the nature of specifications.

What, then, constitutes the principles of specification writing? Basically, the principles of specification writing should encompass those factors which permit the architect or engineer to understand more clearly the relationship between drawings and specifications, between the graphic and the verbal, and to enable him to communicate more effectively by setting forth in a logical, orderly sequence the material to be incorporated within a specification.

## Principles

In broad terms, the principles of specification writing can be set forth as follows:

1. *The Role of the Specifications.* Specifications constitute one of the contract documents, together with the drawings and the agreement. Since they are written instructions, they are frequently adjudged by the courts as having greater importance than drawings

*1*

when these documents are in conflict, and judgments are frequently resolved on the basis of the specifications. Also, the drawings, except for structural, mechanical, and electrical drawings, make no attempt at segregating the work of the various trades, and all of the architectural work is shown on them as an integrated whole. The specifications, on the other hand, segregate the information depicted on the drawings into the various specification sections so that a contractor can generally let his subcontracts on the basis of the specification breakdown of sections.

2. *The Relationship between Drawings and Specifications.* Drawings are a graphic portrayal of the various elements. Specifications should describe the quality of materials, processes, and workmanship. There should not be duplication between these two documents; instead, they should be complementary. To improve coordination between drawings and specifications, there should be standardization of the information appearing in them.

3. *The Organization of Specifications.* For many years specifications were arranged in a series of sections based on the order or chronology in which the various trades appeared on the construction scene. However, it was found that our increasingly complex building structures did not necessarily follow these simple rules, nor was there a uniform, nationwide system of specifications. The *CSI Format for Construction Specifications* has established a uniform arrangement of division-section organization.

4. *The Technical Section and Its Arrangement.* The technical section, which generally forms a subcontract, must be defined in terms of its scope and content. Until the recently promulgated *CSI 3—Part Section Format,* there was no universal arrangement of information in an orderly, coherent series of paragraphs dealing with the content of the technical section.

5. *Types of Specifications.* Specifications can be prepared on the basis of either methods or results. The specifier can elect to specify in detail the method by which a contractor does certain operations in order to achieve a certain result. Conversely, he can prepare a specification placing on the contractor the responsibility for securing the desired result, leaving to the contractor the method by which he secures it. Generally, there are four different types of specifications: descriptive, performance, proprietary, and reference.

6. *Specification Writing Techniques.* These techniques involve the use of scope clauses, the work of other sections, the use of "or equal" or base bid specifications, the avoidance of duplication and repetition, and the use of the residual legatee technique.

## Procedures

In broad terms, the procedures to be followed are based on standards developed by the American Institute of Architects (AIA), the Construction Specifications Institute (CSI), and by systems developed by the author and by others from whom he has borrowed heavily. These include the following:

1. *General Requirements.* These are nonlegal, nontechnical portions of the specifications which are described in detail in Division 1 of the CSI Format and the *Uniform Construction Index.*

2. *Specifying Materials.* This procedure deals with the approach to writing open, closed, or base bid specifications for materials and products, citing the advantages and disadvantages of each system.

3. *Specification Language.* The use of clear technical language that can be understood by contractors, superintendents, and foremen is imperative. Legal phraseology or highly stilted formal terms and sentences are to be avoided. Sentences should be clear and concise; they should be written in simple terms to avoid misunderstanding. Sentence structure, punctuation, and the phraseology used in specification writing are an art in themselves.

4. *Specification Reference Sources.* Knowing where to look for information to be used in specifications is quite important. Materials standards have been established by the Federal Government, the American Society for Testing and Materials, the American National Standards Institute, and others. Association standards have been developed by materials manufacturers and subcontractors for materials and workmanship. Many textbooks on specification writing are available for reference purposes. Guide specifications are available from the American Institute of Architects, and several specification studies are available from Construction Specifications Institute.

5. *Materials Evaluation.* A systematic approach to the evaluation

of materials is outlined suggesting the parameters to review in evaluating and selecting materials.

6. *Specification Writing Procedures.*  A guide is recommended for the procedure to be used in gathering information, research, and writing to dovetail the completion of the specifications with the finalization of the drawings.

This book is intended to be an extension of the principles and procedures above, and it is designed for the student as well as those currently engaged in writing specifications.

# 1

## The Role of the Specifications

Whenever an architect or an engineer is commissioned by an owner to design a building or a structure, he must develop three basic documents which a third party, the contractor, must utilize when he undertakes to build the structure. These three basic documents are the drawings, the conditions of the contract, and the specifications; together with the agreement, they constitute the contract documents. The American Institute of Architects classifies specifications as one of the contract documents—one of the necessary constituent elements of the contract. As one of the major contract documents, it is imperative that practicing architects and engineers have a very good working knowledge of the role that specifications play.

Whether the specifications are written by a specifier in a large office or by the job captain or architect and engineer in a small office, they are utilized by a rather diverse group of participants. To begin with, they are written for the contractor to tell him how to construct, manage, and direct the construction. They are also written for the estimator in the contractor's office, who prepares the estimate based on the specifications. They are written for the purchasing agent in the contractor's office, who procures the materials and equipment described in the specifications. They are written for the resident project representative or inspector, who must be given a document that can aid him in inspecting and controlling the work. They are written for the owner, who would like to know what he is buying and what he is entitled to receive. They are written for the subcontractors so that each can readily discern the scope of his subcontract. They are written for the manufacturers of building materials and equipment so that the grade and type are clearly defined with respect to the many variations they may manufacture.

*Webster's Unabridged Dictionary* gives the following definition of the term "specifications": "Specifications (usually plural)—A written or printed description of work to be done, forming part of the contract and describing qualities of material and mode of construction, and also giving dimensions and other information not shown in the drawings." But the dictionary description does not suffice. As we explore the full meaning of the term, we discover many areas solely within the province of the specifications that extend far beyond a mere elaboration of the drawings.

For example, the specifications alone, as a contract document prepared by the architect, set forth legal requirements, insurance requirements, bidding procedures, alternatives, options, subcontractor limits, and inspection and testing procedures. In many instances, design decisions cannot be shown on the drawings, and the specifications are the only vehicle through which these design considerations can be transmitted to the contractor. The following list illustrates the function of the specifications:

1. *Legal Considerations*
   a. As a written document the courts have generally held that in the event of conflict between drawings and specifications, the specifications govern and judgments are most frequently resolved on the basis of the specification requirements.
   b. General conditions, whether they consist of AIA standard preprinted forms, federal, state, or municipal forms, the Consulting Engineers Council forms, or individually prepared general conditions, are usually bound with the specifications and, by reference, made a part of the specifications. The content and role of the general conditions are elaborated on separately. Essentially, however, they establish the legal rights, responsibilities, and relationships of the parties to the contract.
2. *Insurance Considerations*
   Insurance requirements governing workmen's compensation, con-

tractor's liability, and fire insurance are usually incorporated in the general conditions or in supplementary conditions and, again, made a part of the specifications by incorporation therein.

3. *Bidding Requirements*

The bidding requirements include the Invitation to Bid, the Instructions to Bidders, the Bid Form, and the Bid Bond. These bidding requirements are developed by the architect solely for the use of the bidder and are intended to provide the bidder with information required to submit a proposal. These are usually bound with the specifications and made a part thereof.

4. *Alternatives, Options*

a. The specifications provide a basis for the contractor's estimate and the submission of a bid. Alternatives are established by the architect and owner for the deletion of work, the addition of work, and for the substitution of materials. Alternatives are listed in the Bid Form or Form of Proposal.

b. The technical specifications may permit the contractor, at his option, to use one of several materials or manufacturers' brands specified for use in the work.

5. *Subcontractor's Limits*

Drawings generally show all of the work to be done and the inter-relationship of the various parts. No attempt is made on the drawings to segregate the work of the several subcontractors, except that separate drawings are generally prepared for plumbing, heating, ventilation and air conditioning, and electrical work. The specifications segregate the work shown on the drawings into many sections, or units of work, so that the general contractor can sublet the work to various subcontractors.

6. *Inspection and Testing Procedures (Quality Control)*

The specifications establish inspection and testing procedures to be followed during the construction operations. Standards for office and field inspection are described for numerous materials and building systems. Test procedures are given for evaluating the performance of completed mechanical installations.

7. *Design Criteria*

In some instances the drawings cannot be utilized to show or delineate design decisions. For example, the architect's selection of finish hardware for doors can be described only in the specifications. Specifications for paint materials, the number of coats of paint, and the degree of luster or sheen are similarly given only in the specifications.

## Project Manual

Everyone associated with the design profession (architects, engineers, and specifiers), as well as those involved in construction (contractors, subcontractors, and materials manufacturers), use the term "specifications" when referring to the written document that accompanies drawings. The definition has prevailed for years, even though this particular book contains some documents that cannot be strictly classified as specifications.

Some specifiers say that specifications are only the technical sections. Others state that the specifications constitute everything between the two covers of a book. The material usually bound in a book includes an Invitation to Bid, Instructions to Bidders, a Bid Form (or Proposal Form), a standard preprinted form of general conditions, supplementary conditions, a form of agreement, forms for Bid Bonds, Payment Bonds, and Labor and Materials Bonds.

The inability to define specifications properly lies both in the failure to define many of the documents used in construction and in the absence of any authoritative source establishing precise definitions. The terms "construction documents" and "contract documents" are sometimes used interchangeably. Although contract documents are defined in the *AIA General Conditions,* a definition for construction documents is nonexistent. The term "bidding documents" has been used rather loosely in the past. Some have employed it to mean the drawings and specifications available to bidders in preparing a bid; others have used it to mean the bidding requirements.

The bidding requirements are now defined by both AIA and CSI as including the Invitation to Bid, the Instructions to Bidders, the Bid Form or Proposal Form, together with certain sample forms such as Bid Bond, Performance and Payment Bonds, and similar documents.

The agreement on the definition of bidding requirements resolved somewhat the proper terms to be used for the parts that constitute these documents. Advertisement to Bid, Notice to Bidders, and Notification to Contractors have been used in place of the recently adopted term "Invitation to Bid." Other terms used for Instructions to Bidders have included Information for Bidders and Conditions of Bid. The terms "Bid Form" and "Proposal Form" have also been used extensively in the past, and agreement has not been reached on a single term. CSI documents call it Proposal Form, and AIA documents call it Bid Form.

Confronted by this profusion of terms, the profession is slowly making progress in redefining some documents. In an attempt to clarify the various documents prepared by architects for detailing, specifying, bidding, and constructing a project, the AIA, through a national Committee on Specifications in 1965, produced the "Project Manual Concept." The chapter on specifications in the *AIA Handbook of Professional Practice* was updated to include a reorganization and renaming of the old specifications, and it provides a new concept —the "Project Manual." The AIA describes its content and function. The following has been reproduced with the permission of the American Institute of Architects. Further reproduction is not authorized.

"The PROJECT MANUAL concept is, in its simplest terms, a reorganization and renaming of that familiar book of bidding forms and contract documents, usually referred to as the "Specifications" or "Specs" which, along with the drawings, are the documentary basis for all construction projects. The following Outline of Contents is an updating of the PROJECT MANUAL/format first published in the November 1965 AIA JOURNAL.

"The PROJECT MANUAL contains a great deal more than specifications. It normally includes the bidding documents; i.e., invitation, instructions, sample bid, bond and agreement forms, general and supplementary conditions, and information on alternate and unit prices, in addition to the *technical specifications* describing the materials and performance expected in the construction of the project. The book also frequently contains a schedule of the drawings pertaining to the Project. The book is indeed a *manual of project bidding* requirements and contract documents.

"Prior to 1960 many architects used the ADVERTISEMENT, INSTRUCTION TO BIDDERS and SPECIFICATIONS rather loosely. In many cases, requirements of the contract were placed in the INSTRUCTION TO BIDDERS; when serious problems required resolution in court, attorneys could raise the question whether information in these other documents was actually a part of the contract requirements. During the 1960's, various AIA committees working together and with other professional societies established criteria for organization of this material which would include all contract requirements in the *specifications* and locate all information and material that would become void after contracts with the earlier documents. This was confirmed in the 1966 edition of the General Conditions which stated in positive terms that the Contract superceded all prior negotiations or agreements.

"After considerable discussion, however, it was conceded that for convenience these earlier documents, which were actually not a part of the contract, should be bound with the specifications into a PROJECT MANUAL.

"The material included in the PROJECT MANUAL fall into two general categories: 1) those describing the requirements for bidding and 2) those that become part of the contract documents upon the signing of the construction contract. Within each of these two categories, all of the familiar instructions, forms, and the like are organized as outlined here.

"The AIA Board in adopting this concept instructed the various committees within the Commission on Professional Practice to proceed with the detailing of consequent changes in all AIA office practice documents. All of the current AIA documents now have been edited to work with the PROJECT MANUAL concept of organization.

"By the use of this system, reference can be made to the four parts of a contract: 1) Agreement, 2) Conditions of the Contract, 3) Schedule of Drawings and 4) Specifications. All of these are in the PROJECT MANUAL except the drawings, and it is recommended that a list of the drawings be included for information. Neither the INVITATION TO BID, INSTRUCTION TO BIDDERS or SAMPLE FORMS are in the contract, but are bound into the PROJECT MANUAL for the convenience of the bidders."

The sequence recommended by the AIA for the material to be bound in the Project Manual is as follows:

Title Page
Table of Contents
Addenda (if bound in Project Manual)
1.0  Bidding Requirements
  The bidding requirements are bound into the Project Manual
  with the contract documents for the convenience of the bidders.
1.1  Invitation to Bid or Advertisement for Bids
  - Exact title of project and its location
  - Name of owner
  - Name of architect
  - Person to receive bids
  - Place for receipt of bids
  - Time for receipt of bids
  - Type of bid opening
  - Short description of project, scope, and type of construction
  - Type of contract
  - Place for examining bidding documents
  - Place for obtaining bidding documents
  - Time for bidding documents to be available
  - Procedure for obtaining bidding documents
  - Statement of what bonds are required
  - Statement on time of completion and liquidated damages, if any
1.2  Instructions to Bidders
  - Qualifications of bidders
  - Bidder's representation
  - Examination of bidding documents
  - Clarification of bidder's questions
  - Addenda
  - Bid guarantee requirements (Bid Bond)
  - Performance Bond and Labor and Materials Payment Bond
  - Substitutions
  - Procedure for execution of bids
  - Procedure for submission of bids
  - Procedure for withdrawal or modification of bids
  - Procedure for opening of bids
  - Conditions for rejection of bids
  - Procedure for award of contract
  - Submission of post-bid information

- Return of bidding documents
- Other instructions to bidders

1.3 Sample Forms
- Bid
- Bid Bond
- Power-of-attorney
- Bidder's Qualification Questionnaire
- Agreement form
- Performance and Payment Bonds
- Noncollusion affidavit
- Certificate of insurance
- Consent of surety
- Application and certificate for payment
- Other sample forms

2.1 Agreement
2.2 Conditions of the Contract
   2.2.1 General Conditions
- AIA Document A201
   2.2.2 Supplementary Conditions
- Examination of site
- Labor standards
- Wages and hours
- Insurance requirements
- Unit prices, predetermined
- Payment to the contractor
- Time of completion
- Partial occupancy
- Bonus and penalty clause
- Liquidated damages
- Guarantees and affidavits
- Type of contract (single or separate)
- Substitution of materials
- Other conditions as required

2.3 Schedule of Drawings
2.4 Specifications
   2.4.1 General Requirements
- Allowances
- Summary of the work

- Alternates
- Submittals
- Schedules and reports
- Samples and shop drawings
- Temporary facilities
- Quality controls
- Project meetings
- Cleaning up
- Project closeout

2.4.2   Site Work
2.4.3   Concrete
2.4.4   Masonry
2.4.5   Metals
2.4.6   Wood and Plastics
2.4.7   Thermal and Moisture Protection
2.4.8   Doors and Windows
2.4.9   Finishes
2.4.10  Specialties
2.4.11  Equipment
2.4.12  Furnishings
2.4.13  Special Construction
2.4.14  Conveying Systems
2.4.15  Mechanical
2.4.16  Electrical End of Copyright Material.

The term "specifications" has been used for a long time to describe the bound volume, and many specifiers will be loathe to change, or to use the new term. We should be realistic, however, and recognize that some of the documents bound in the old familiar volume are not specifications, and that we cannot continue to refer to this volume as such.

# 2

## *Relationship between Drawings and Specifications*

The information that is necessary for the construction of any structure is usually developed by the architect by means of two basic documents: the drawings and the specifications. These two documents represent a means of communication of information between architect and contractor, but each document uses a special form of communication: one pictorial and the other verbal. Yet, in spite of these distinct methods of transmitting information, the documents should complement one another, and neither should overlap nor duplicate. In this way, each document fulfills its own function. In broad terms, the drawings are a graphical portrayal and the specifications are a written description of the legal and technical requirements forming the contract documents. Each should convey its own part of the story completely, and neither should repeat any part that properly belongs to the other, since duplication can very often result in differences of meaning.

Drawings present a picture, or a series of pictures, of the structure or parts of a structure to be erected. They give the size, form, location, and arrangement of the various elements. This information cannot be described in the specifications since it is graphically shown by means of lines, dots, and symbols peculiar to drawings. In fact, a drawing is a special language or means of communication to convey ideas of construction from one person to another. These ideas cannot be conveyed by the use of words.

Specifications are, by their very nature, a device for organizing the information depicted on the drawings. The drawings show the interrelationship of all the parts that go together to make the grand design. It has only been a relatively short time since mechanical, electrical, and structural information have been shown on separate drawings. All the general construction details are shown on drawings as they relate to one another, with no attempt to separate diverse materials. It is the specifications that break down the interrelated information shown on drawings into separate, organized, and orderly units of work, which we refer to as technical sections of the specifications.

To maintain the separate yet complementary character of these two documents and to ensure that they will be interlocking but not overlapping requires the development of definite systems for each. Hence what is better described in the specifications should not be shown on the drawings and, similarly, what is better shown on the drawings should not be described in the specifications.

Drawings should generally show the following information:

1. Location of materials, equipment, and fixtures.
2. Detail and overall dimensions.
3. Interrelation of materials, equipment, and space.
4. Schedules of finishes, windows, and doors.
5. Sizes of equipment.
6. Identification of class of material at its location.
7. Alternatives.

Specifications should generally describe the following items:

1. Type and quality of materials, equipment, and fixtures.
2. Quality of workmanship.
3. Methods of fabrication, installation, and erection.
4. Test and code requirements.
5. Gauges of manufacturers' equipment.
6. Allowances and units.
7. Alternatives and options.

Specifications should not overlap or duplicate information contained on the drawings. Duplication, unless it is repeated exactly word for word, is harmful because it can lead to contradiction, confusion, misunderstanding, and difference of opinion. Duplication, word for word, is redundant.

To achieve proper separation of information between drawings and specifications, it is essential that the development of the specifications go hand in hand with the preparation of the drawings. At the outset, someone in the office should be made responsible for establishing and keeping the all-important checklist for a specific project. This checklist should establish a schedule of what is to appear on the drawings, what is to be described in the specifications, and what is to be itemized and listed in schedules on the drawings. The checklist should include preliminary or outline specifications, lists of all decisions made in the drafting room, and notes of all changes made on the drawings since the last set was printed for the specifier, including questions to be settled.

The broad guidelines previously noted for the separation of material that appears on the drawings and in the specifications do not go far enough in establishing a line of demarcation between these documents, inasmuch as there are areas of disagreement among authorities on specifications writing as to the specific information that should be shown or specified or both. For example, one authority suggests that the drawings should indicate a material such as concrete and the specifications should determine whether it is to be precast or cast-in-place concrete. It would be preferable that the drawings delineate the location of these two different materials. Another authority argues against the customary hatching and other indication of materials on plans and elevations. If the experts disagree, how can the draftsman and the neophyte specifications writer settle the issue? Duplication will exist between drawings and specifications when there is a lack of a clear-cut and well-defined policy.

Generally, each office establishes a policy to be followed in its own practice. However, systems can be formulated between the specifications writer and the draftsman, and as a general rule it will follow that common sense will dictate which medium serves as the better means of communication.

To ensure complete understanding on the part of the contractor, it is essential that standard terminology be employed and used consistently on both drawings and specifications. Too often draftsmen use certain terms to identify materials on the drawings, which may differ from the terms employed by the specifier. For example, a draftsman may use the term "calking" to describe all calking and sealant work, whereas the specifier will be selective and discriminate between the choice of materials and terms, resulting in ambiguity and misunderstanding on the part of the contractor.

Quite often it is essential to identify classes of materials at specific locations so that the contractor can readily differentiate between the variety of classes of materials. For example, there may be several types of flashings illustrated on the drawings or several varieties of sealants shown. By ascribing numerical or alphabetical characters to these flashings or sealants, both on the drawing and in the specifications, the contractor has no problem identifying what material goes where. This system precludes the necessity for describing in the specifications the location of classes of similar materials.

# 3

## Organization of Specifications

When the term "organization of specifications" is used, it refers to the separation of the specifications into a series or a schedule of separate units of work termed "technical sections." (See Chapter 4 for a definition of the technical section.) The history of specification writing and the growth in complexity of buildings illustrate how this system of a series of technical sections has evolved.

Specifications in the eighteenth and nineteenth centuries consisted of a single document containing a description of all the work and materials to be included in a building. This was especially true of small, simple structures that were constructed by a general contractor who engaged all the crafts and did not sublet or subcontract any parts of the work. An early textbook entitled *Handbook of Specifications* by T. L. Donaldson, London, 1860, provided for the arrangement of specifications on a craft basis. The specifications were divided into two main general divisions, with subdivisions as follows:

| *Carcase* | *Finishing* |
|-----------|-------------|
| Excavator | Joiner |
| Bricklayer | Plasterer |
| Mason | Plumber |
| Slater | Painter |

Founder and smith            Glazier
Carpenter                    Paperhanger
                             Ironmonger
                             Smith and bellhanger
                             Gasfitter

As materials and methods of construction gained a degree of sophistication, the specification sections began to change with the times and took on additional crafts within one section of the specifications. In lieu of the general contractor hiring specific crafts under his own supervision, he began to sublet portions of the work to subcontractors, who, in turn, hired the several crafts to perform certain parts of the work.

Specifications written for buildings toward the end of the nineteenth century consisted generally of three main sections—masonry, carpentry, and mechanical work—with various allied or related subjects under each section. The masonry section included excavation, concrete, brickwork, stonework, steel columns and lintels, and waterproofing. The carpentry section included roofing, glazing, and painting, as well as carpentry. The mechanical or pipe trades consisted of plumbing, gas, and heating work. When electricity came into use, it was included in the mechanical work.

The foregoing arrangement of the specifications was adequate for the nature of the buildings constructed at that time, and for the materials and methods of construction prevalent at that time. However, buildings grew more complex, and materials and construction techniques became more involved. It became necessary to increase the number of technical sections as more portions of the work came under subcontract. Today, the specification sections are designed essentially to permit general contractors, estimators, subcontractors, manufacturers, and materials dealers to "take off" the items of their specific work for estimate during the bidding period.

Accuracy in estimating is in the best interests of building owners and architects alike. To assure accuracy, the specifications should be divided into sections to permit the contractor's estimator and the subcontractors to prepare estimates quickly and precisely. The preparation of drawings and specifications takes considerable time, whereas bidding periods are generally of short duration. It is therefore quite evident that the estimator must have a specification separated by sections or units of work so that he can list the materials and quantities, note the methods of their use and installation, separate those parts on which he will take subestimates, secure prices and tabulate results, all within a three- or four-week bidding period, and often within only two weeks. Such a system also permits the specification writer to organize his own material. It provides him with a method for organizing the information on the drawings in a systematic, orderly, and prearranged manner.

For convenience in writing, speed in estimating, and ease of reference, it has been found that the most suitable  organization of the specifications is a series of sections dealing successively with the different subcontractors, and in each section grouping all of the work of the particular trade to which the section is devoted.

Until April 1963, when the *CSI Format for Construction Specifications* was promulgated by the Construction Specifications Institute, each specification writer organized his specifications in a series of sections which more or less followed a time relationship, or chronological order, related to the order of appearance on the site of the various subcontractors. However, from office to office, and even within the same office, this order was not uniform. In addition, complex structures required that certain mechanical trades be involved at an early stage in the construction process so that a true trades-chronology was not possible in the organization of the specifications. It became apparent that a major overhaul was required in the organization of specification sections, and that a uniform system would provide certain corollary benefits.

The *CSI Format for Construction Specifications* is reproduced as follows:

# CSI FORMAT

## DIVISION 1—GENERAL REQUIREMENTS
01010 SUMMARY OF WORK
01100 ALTERNATIVES
01200 PROJECT MEETINGS
01300 SUBMITTALS
01400 QUALITY CONTROL
01500 TEMPORARY FACILITIES & CONTROLS
01600 PRODUCTS
01700 PROJECT CLOSEOUT

## DIVISION 2—SITE WORK
02010 SUBSURFACE EXPLORATION
02100 CLEARING
02110 DEMOLITION
02200 EARTHWORK
02250 SOIL TREATMENT
02300 PILE FOUNDATIONS
02350 CAISSONS
02400 SHORING
02500 SITE DRAINAGE
02550 SITE UTILITIES
02600 PAVING & SURFACING
02700 SITE IMPROVEMENTS
02800 LANDSCAPING
02850 RAILROAD WORK
02900 MARINE WORK
02950 TUNNELING

## DIVISION 3—CONCRETE
03100 CONCRETE FORMWORK
03150 EXPANSION & CONTRACTION JOINTS
03200 CONCRETE REINFORCEMENT
03300 CAST-IN-PLACE CONCRETE
03350 SPECIALLY FINISHED CONCRETE
03360 SPECIALLY PLACED CONCRETE
03400 PRECAST CONCRETE
03500 CEMENTITIOUS DECKS

## DIVISION 4—MASONRY
04100 MORTAR
04150 MASONRY ACCESSORIES
04200 UNIT MASONRY
04400 STONE
04500 MASONRY RESTORATION & CLEANING
04550 REFRACTORIES

## DIVISION 5—METALS
05100 STRUCTURAL METAL FRAMING
05200 METAL JOISTS
05300 METAL DECKING
05400 LIGHTGAGE METAL FRAMING
05500 METAL FABRICATIONS
05700 ORNAMENTAL METAL
05800 EXPANSION CONTROL

## DIVISION 6—WOOD & PLASTICS
06100 ROUGH CARPENTRY
06130 HEAVY TIMBER CONSTRUCTION
06150 TRESTLES
06170 PREFABRICATED STRUCTURAL WOOD
06200 FINISH CARPENTRY
06300 WOOD TREATMENT
06400 ARCHITECTURAL WOODWORK
06500 PREFABRICATED STRUCTURAL PLASTICS
06600 PLASTIC FABRICATIONS

## DIVISION 7—THERMAL & MOISTURE PROTECTION
07100 WATERPROOFING
07150 DAMPPROOFING
07200 INSULATION
07300 SHINGLES & ROOFING TILES
07400 PREFORMED ROOFING & SIDING
07500 MEMBRANE ROOFING
07570 TRAFFIC TOPPING
07600 FLASHING & SHEET METAL
07800 ROOF ACCESSORIES
07900 SEALANTS

## CSI FORMAT

### DIVISION 8—DOORS & WINDOWS

08100 METAL DOORS & FRAMES
08200 WOOD & PLASTIC DOORS
08300 SPECIAL DOORS
08400 ENTRANCES & STOREFRONTS
08500 METAL WINDOWS
08600 WOOD & PLASTIC WINDOWS
08650 SPECIAL WINDOWS
08700 HARDWARE & SPECIALTIES
08800 GLAZING
08900 WINDOW WALLS/CURTAIN WALLS

### DIVISION 9—FINISHES

09100 LATH & PLASTER
09250 GYPSUM WALLBOARD
09300 TILE
09400 TERRAZZO
09500 ACOUSTICAL TREATMENT
09540 CEILING SUSPENSION SYSTEMS
09550 WOOD FLOORING
09650 RESILIENT FLOORING
09680 CARPETING
09700 SPECIAL FLOORING
09760 FLOOR TREATMENT
09800 SPECIAL COATINGS
09900 PAINTING
09950 WALL COVERING

### DIVISION 10—SPECIALTIES

10100 CHALKBOARDS & TACKBOARDS
10150 COMPARTMENTS & CUBICLES
10200 LOUVERS & VENTS
10240 GRILLES & SCREENS
10260 WALL & CORNER GUARDS
10270 ACCESS FLOORING
10280 SPECIALTY MODULES
10290 PEST CONTROL
10300 FIREPLACES
10350 FLAGPOLES
10400 IDENTIFYING DEVICES
10450 PEDESTRIAN CONTROL DEVICES
10500 LOCKERS
10530 PROTECTIVE COVERS
10550 POSTAL SPECIALTIES
10600 PARTITIONS
10650 SCALES
10670 STORAGE SHELVING
10700 SUN CONTROL DEVICES (EXTERIOR)
10750 TELEPHONE ENCLOSURES
10800 TOILET & BATH ACCESSORIES
10900 WARDROBE SPECIALTIES

### DIVISION 11—EQUIPMENT

11050 BUILT-IN MAINTENANCE EQUIPMENT
11100 BANK & VAULT EQUIPMENT
11150 COMMERCIAL EQUIPMENT
11170 CHECKROOM EQUIPMENT
11180 DARKROOM EQUIPMENT
11200 ECCLESIASTICAL EQUIPMENT
11300 EDUCATIONAL EQUIPMENT
11400 FOOD SERVICE EQUIPMENT
11480 VENDING EQUIPMENT
11500 ATHLETIC EQUIPMENT
11550 INDUSTRIAL EQUIPMENT
11600 LABORATORY EQUIPMENT
11630 LAUNDRY EQUIPMENT
11650 LIBRARY EQUIPMENT
11700 MEDICAL EQUIPMENT
11800 MORTUARY EQUIPMENT
11830 MUSICAL EQUIPMENT
11850 PARKING EQUIPMENT
11860 WASTE HANDLING EQUIPMENT
11870 LOADING DOCK EQUIPMENT
11880 DETENTION EQUIPMENT
11900 RESIDENTIAL EQUIPMENT
11970 THEATER & STAGE EQUIPMENT
11990 REGISTRATION EQUIPMENT

## CSI FORMAT

**DIVISION 12—FURNISHINGS**
12100 ARTWORK
12300 CABINETS & STORAGE
12500 WINDOW TREATMENT
12550 FABRICS
12600 FURNITURE
12670 RUGS & MATS
12700 SEATING
12800 FURNISHING ACCESSORIES

**DIVISION 13—SPECIAL CONSTRUCTION**
13010 AIR SUPPORTED STRUCTURES
13050 INTEGRATED ASSEMBLIES
13100 AUDIOMETRIC ROOM
13250 CLEAN ROOM
13350 HYPERBARIC ROOM
13400 INCINERATORS
13440 INSTRUMENTATION
13450 INSULATED ROOM
13500 INTEGRATED CEILING
13540 NUCLEAR REACTORS
13550 OBSERVATORY
13600 PREFABRICATED BUILDINGS
13700 SPECIAL PURPOSE ROOMS & BUILDINGS
13750 RADIATION PROTECTION
13770 SOUND & VIBRATION CONTROL
13800 VAULTS
13850 SWIMMING POOLS

**DIVISION 14—CONVEYING SYSTEMS**
14100 DUMBWAITERS
14200 ELEVATORS
14300 HOISTS & CRANES

14400 LIFTS
14500 MATERIAL HANDLING SYSTEMS
14570 TURNTABLES
14600 MOVING STAIRS & WALKS
14700 PNEUMATIC TUBE SYSTEMS
14800 POWERED SCAFFOLDING

**DIVISION 15—MECHANICAL**
15010 GENERAL PROVISIONS
15050 BASIC MATERIALS & METHODS
15180 INSULATION
15200 WATER SUPPLY & TREATMENT
15300 WASTE WATER DISPOSAL & TREATMENT
15400 PLUMBING
15500 FIRE PROTECTION
15600 POWER OR HEAT GENERATION
15650 REFRIGERATION
15700 LIQUID HEAT TRANSFER
15800 AIR DISTRIBUTION
15900 CONTROLS & INSTRUMENTATION

**DIVISION 16—ELECTRICAL**
16010 GENERAL PROVISIONS
16100 BASIC MATERIALS & METHODS
16200 POWER GENERATION
16300 POWER TRANSMISSION
16400 SERVICE & DISTRIBUTION
16500 LIGHTING
16600 SPECIAL SYSTEMS
16700 COMMUNICATIONS
16850 HEATING & COOLING
16900 CONTROLS & INSTRUMENTATION

*a* This document has been reproduced with the permission of the Construction Specifications Institute. Further reproduction is not authorized.

How does one use this new system? For each specific project the specification writer prepares his technical sections as he did previously, except that he now places them under the fixed division.

Where local trade practices or conditions of the specific project dictate, the specification writer has the prerogative to alter the location of the information and the section.

The *CSI Format for Construction Specifications,* now much more updated by the *Uniform Construction Index,* is somewhat analogous to the organization of specifications referred to earlier in the *Handbook of Specifications.* The division headings under the CSI Format are based on four major categories: materials, trades, functions of work, and place relationships. For example, Division 4, Masonry, is an instance of a materials relationship. The sections listed under Division 4 have materials as their common denominator. These include sections on unit masonry, stone, and mortar and masonry restoration, all dealing with materials common to one another. Division 5, Metals, is an example of a trades relationship. Sections on structural metal, open-web joists, metal decking, miscellaneous metal, and ornamtntal metal are located here, and the metal fabricators, erectors, and ironworks usually perform this type of work. Division 7, Thermal and Moisture Protection, is illustrative of a relationship based on functions of work. The sections dealing with the environmental protection of the building are located here: roofing, waterproofing, dampproofing, thermal insulation, and caulking and sealing. Division 2, Site Work, is an example of a place relationship, and it includes such sections as demolition, clearing and grubbing, earthwork, piling, roads and walks, and lawns and planting.

What the the advantages of this particular system for organizing the specifications? In preparing his specification section, the specifier no longer needs to be concerned with whether the architect or the engineer adds or deletes certain materials or trades as he develops his drawings. Previously, this change in design meant the deletion of a specification section or the inclusion at the last moment of a specification section, placed entirely out of sequence. Under the *Uniform Construction Index* such revisions do not impose hardships since a section can be added to or deleted from a specific division without radically upsetting the sequence and numbering system. The specifier can write or prepare sections long before the drawings are completed and assign numbers to them immediately. In addition, the specification writer can now file material, shop drawings, correspondence, technical data, literature, samples, estimates, and a host of office memoranda under a similar numbering system. The contractor, manufacturer, estimator, and inspector can more readily find those items in the specifications with which he is concerned.

In an earlier version of this book the author predicted that in time, with widespread use, the AIA would conform its specification work sheets and its standard filing system to this system; the Associated General Contractors could number its estimating work sheets on the same basis; Sweet's Catalog Service could renumber its architectural file; and building materials manufacturers could number their literature accordingly. This prediction has been fulfilled. (See Chapter 19 for a more complete review of the interrelationship of specification sections, data and literature filing, and contractors' estimates.)

# 4

## Concept of the Technical Section

One should visualize a complete specification as a series of chapters in a novel, with each section of a specification comparable to the chapter in that novel. The primary reason for establishing chapters in a novel is basically to make it easer to read and to provide an underlying theme in that chapter. The specification section is also a subdivision of the complete specification, providing information on one subject as the content of each technical section.

A technical section in its simplest term is a word description of a basic trade or material installation, outlining the quality of material to be used and the quality of workmanship to be employed in its installation. A technical section can best be described as a unit of work consisting of a carefully worded description of materials and an explanation of methods of construction in the form of instructions to a contractor. The term "trade section" has been used frequently heretofore to describe this unit; however, under today's connotations and definitions, trade section is misleading, and much misunderstanding has arisen from the use of the term as it relates to the technical sections of specifications.

It has previously been pointed out that the drawings generally show all the work that is to be constructed. The only attempt made in the drawings to segregate the work of different trades is in the preparation of separate drawings for plumbing, heating, electrical work, and structural work. When a specification is written, the specifier endeavors to segregate under the various technical or trade sections of the specifications a unit of work that a materials man may supply for another to install, a unit of work that combines the responsibilities of several subcontractors into a single authority, or a unit of work that is performed by a single recognized trade.

The misunderstanding concerning the use of the word "trade" arises both from the dictionary definition and from the failure to recognize that the so-called trade section of the specifications can be as restrictive or all-inclusive as previously described. The dictionary defines trade as: "(1) The business one practices or the work in which one engages regularly; occupation; means of livelihood. (2) A pursuit requiring manual or mechanical training and dexterity; a craft. (3) Those engaged in a business or industry." Trade can therefore mean a craft, such as carpentry, bricklaying, or plumbing; or it can mean a business, such as a concrete subcontractor or a plumbing and heating subcontratcor.

A unit of work that a contractor may let to a subcontractor can encompass a section entitled "Concrete Work." The concrete subcontractor employs carpenters for erecting formwork; lathers or ironworkers for installing the steel reinforcement; concrete laborers for placing concrete; and cement masons for finishing the concrete.

A unit of work that a materials man may supply for another to install is exemplified by the section entitled "Finish Hardware." The general contractor purchases the hardware from a materials supplier, who simply delivers the material to the site for the carpentry subcontractor to install, or for the general contractor to install with his own forces.

A unit of work that combines several subcontractors so that a single responsibility is established for that portion of the work is illustrated by the section entitled "Curtain Walls." The general contractor may award this work to one subcontractor, who in turn will sublet such items of work as fabrication and installation of metal framing and metal panels, furnishing and installation of glass, caulking and sealing of the curtain wall, flashing of the curtain wall, and insulation of the curtain wall.

A unit of work that is performed by a single recognized trade can best be illustrated by the section entitled "Painting." While the work can be done by a subcontractor who employs only painters, the general contractor may elect to hire his own painting crew to perform this portion of the work.

It is not a simple matter to determine the proper subdivision of the technical sections, and once made it is not necessarily permanent. Changes occur as new materials are introduced by building materials manufacturers, and recognized trades change as a result of these new materials. Changes also occur as new concepts in design appear, such as the curtain wall and the integrated ceiling. Changes are also dictated by the introduction of new construction techniques, such as lift slabs and slip forming. Concrete work was formerly a general mason's work; now it is specified under the concrete division and performed by a concrete subcontractor. Wood forms for concrete work were once specified under carpentry, but are now specified under concrete work. Doors that were traditionally wood were installed by carpenters and specified under the carpentry section. When tempered glass doors, metal doors, and bronze doors were introduced, the carpenters claimed this work as being under their jurisdiction. As new methods of work develop, they are first performed by an existing craft, but eventually come under the jurisdiction of specialty subcontractors, where new skills must be developed.

The technical section can vary in size or scope according to the specific project. The concrete section for a nonfireproof dwelling will be small in scope and the descriptive material brief; however, it will still be necessary to write this section as a unit of work to be let as a subcontract. For a high-rise fireproof structure, the concrete section will encompass many aspects of concrete work—admixtures, testing, hot and cold weather concreting, concrete finishes, and form removal—and this technical section will be relatively long and involved. There is, however, another criterion to be considered in determining whether a long, involved concrete section should be written as a single unit of work. Is the amount of work so large in scope and its dollar value so high that it becomes too unmanageable and out of the reach of a subcontractor? In this case, consideration should be given to establishing several technical sections involving certain units of work that may be subcontracted. The concrete reinforcement in a very large project can be established as a separate technical section. In addition, concrete-form work, concrete testing, and the purchase of ready-mix concrete can be established as separate units of work within their respective technical sections.

To establish the technical sections for a project, the specifier should assume that the general contractor may desire to sublet all the work, doing nothing himself except to organize and manage the project. If the specification sections are planned on this basis, it will be possible for the general contractor to reserve for himself whatever parts of the work he may be equipped to do, and sublet all the other parts. It would be quite simple for him to sublet two or more sections to one subcontractor if the work is broken down into small units of work, but it would be difficult for him to divide certain parts of the work between two subcontractors if these parts were not properly separated in the specifications.

It is generally true that a large number of smaller units of work simplifies the work of the estimator, makes it easier for the superintendent to refer to the specifications for any particular part of the work, and aids the specifier in his note-taking for writing the section. It is also necessary to point up the fact that there are certain instances where the work is so closely united in execution as to be combined in one technical section to simplify handling and to place a combination of two or more subcontracts on the shoulders of one subcontractor, who may then sublet part of the work. The curtain wall, the integrated ceiling, and roofing and sheet metal work are examples of composite construction that dictate broader technical sections.

However, the specifier should not lose sight of the fact that although he may establish the technical sections with the scope of each as limited or as broad as he may elect, it is still the privilege of the general contrctor to combine or distribute the various technical sections in any manner he wishes, or to use them and let them as written. The general conditions of the contract should be implemented with the following admonition:

"The following technical sections are generally divided
into units of work for the purpose of ready reference.

The division of the work among his subcontractors is the Contractor's responsibility and the architect assumes no responsibility to act as arbiter to establish subcontract limits between any sections of the work."

It should be apparent by now that the scope, content, and nature of the technical section must be flexible. Sections within the same project can be long and involved, yet others may be short and still represent a large percentage of the work. Conversely, their length and content can vary from project to project, and fixed rules for size and content cannot be established. The specifier must thus assess each project on the merits of its own peculiarities and requirements before establishing the scope and content of the individual technical sections.

In order to arrive at more uniform practices nationwide, the CSI, the AIA, and others have published a list of preferred section titles to be used under the respective division headings of the *Uniform Construction Index*. These section titles have been established on the basis of broad section titles and restrictive section titles, following the concept that a technical section can be written as to be all inclusive or restricted in scope, depending on the particular project. This arrangement provides the flexibility needed by the specifier to retain the prerogative of organizing his specifications, while at the same time establishing a uniform system of preferred section titles throughout the profession. (See Chapter 19).

# 5

## *Arrangement of the Technical Section*

The arrangement of the subject matter in an orderly, comprehensive format within a technical section is important for several reasons. The specifier when following a definitive procedure is less likely to overlook any item. Similarly, the contractor, estimator, materials manufacturer, and inspector will then find the information much more easily in the individual section.

A technical section in a book of specifications can be considered as analogous to a chapter in a book; the chapter, in turn, consists of paragraphs. The material that comprises the section consists essentially of paragraphs and subparagraphs. Other names to describe the breakdown of the material within the technical section, such as articles, clauses, headings, categories, or units, can lead to confusion.

The technical section contains two categories of paragraphs, namely, the technical and nontechnical, as follows:

| *Technical* | *Nontechnical* |
|-------------|----------------|
| Materials | Scope of work |
| Fabrication | Delivery of materials |
| Workmanship | Samples and shop drawings |
| Installation | Permits |
| Tests | Guarantees |
| Schedules | Cleaning |
| Preparation | Job conditions |

In prior years before the advent of a nationally promulgated section format, the technical section was written to include the technical and nontechnical paragraphs in the order in which they occurred chronologically, that is, in a sequence in which the contractor would ordinarily do his work, and each paragraph heading would be simple and self-explanatory. When the specifier followed this course, he was less likely to omit something, and his reliance on a checklist at his side diminished accordingly.

In 1969 CSI developed and promulgated the CSI Section Format. This was refined and updated in August 1970 as CSI Document MP-2B. This nationally approved format provides guidelines for the arrangement of information within a tetchnical section of the specifications, and it offers a concise and orderly method for specifiers to follow.

Prior to the publication of this document, specifiers arranged the information within their technical sections in accordance with their own formulas, and in many instances without any specific method. In many cases, the lack of organization resulted in duplication and omission of information.

The standard CSI Section Format is another important step toward providing a more unified approach. It permits easier access to information by manufacturers, contractors, and inspectors. It provides a checklist for the specifier so that omission of information is minimized. It provides standardization of input that permits its use in connection with computerized specifications and information retrieval.

The CSI Section Format provides for the arrangement and presentation of information under three separate parts as follows:

Part 1, General, is concerned with the ground rules under which the work is to be performed, and it also establishes the scope of work to be performed within the section.

Part 2, Products, is intended for descriptions of materials, equipment and fixtures, and for the manufacturing process used in the development and production of products. The latter requirement includes mixing and fabrication which are inherent in the manufacturing process, whether performed on or off the site. An example of mixing that produces a product off-site is bituminous mixtures for road work. An example of mixing that produces a product on-site is terrazzo. In either case, the development of a product involving mixing is described under Part 2. An example of fabrication is hollow metal doors. The component materials comprising the door and fabrication process are both specified under Part 2 in the description of the development of the basic product.

Part 3, Execution, is used to describe in detail the workmanship, erection, installation, and application procedures.

Under each part the CSI Section Format provides for several standard paragraph headings following a more or less regular sequence or order. Obviously, each paragraph heading may not be pertinent for every technical section, and will not be used where not applicable. In addition, where paragraph heading titles would be forced, they should be retitled to be consonant with the work intended under a specific heading. For example, under Part 3, Execution, there is a paragraph entitled "Installation." Under a section for "Earthwork," this title could be changed to "Excavation." In a section entitled "Structural Steel," the term "Installation" could be changed to "Erection."

The paragraph headings also can take into account certain requirements normally used in civil engineering or heavy construction specifications and in mechanical and electrical specifications. For civil engineers and in heavy construction involved in contracts based on unit prices, a paragraph entitled "Measurement and Payment" can be added to Part 1 which will allow for description of these units. For mechanical and electrical engineers there are paragraphs entitled "Description of Systems" and "Adjusting" that permit the engineer to fully describe involved systems under Part 1 prior to specifying its accomplishment. Specifications for balancing the heating and ventilating system or other mechanical and electrical items can be described under "Adjusting."

The *CSI 3-Part Section Format* has the following broad basic outline:

PART 1   GENERAL
Description
Quality Assurance
Submittals

Product Delivery, Storage and Handling
Job Conditions
Alternatives
Guarantee
PART 2   PRODUCTS
Materials
Mixes
Fabrication and Manufacture
PART 3   EXECUTION
Inspection
Preparation
Installation/Application/Performance
Field Quality Control
Adjust and Clean
Schedules

The major paragraph headings lend themselves in turn to subordinate subparagraphs, which may include the following requirements:

## PART 1   GENERAL

### 1.1  DESCRIPTION

a. *Related Work Specified Elsewhere:* Describe those items that are normally part of this section which the specifier has for one reason or another specified elsewhere. Do not list items that are not normally the work of this section.

b. *Description of System:* Describe systems such as a heating system, cooling system, elevator system, and integrated ceiling system.

### 1.2  QUALITY ASSURANCE

a. *Qualifications:* Establish standards and criteria for determining the qualifications of tradesmen, suppliers, subcontractors, and products.

b. *Requirements of Regulatory Agencies:* Cite the specific needs for permits, codes, ordinances, and UL regulations which the work of this section will require.

---

[a] The extract above has been reproduced with the permission of the Construction Specifications Institute. Further reproduction is not authorized.

c. *Tolerances:* Establish the tolerances of fabrication, manufacture, and erection requirements.

d. *Mock-up:* A mock-up is defined in *Webster's Dictionary* as a "full-sized structural model built accurately to scale chiefly for study, testing or display." Include sample panels, curtain wall assemblies, precast concrete panels, integrated ceiling systems, and so on.

e. *Source Quality Control:* Specify control of products produced off-site at plants, mills, or factories through required testing procedures.

1.3 SUBMITTALS

a. Enumerate the various types of data to be submitted for the architect's review. Assemble the information under various subparagraph headings, such as samples, shop drawings, manufacturers literature, certificates, guarantees, bonds, and so on.

1.4 PRODUCT HANDLING, STORAGE, AND DELIVERY

a. General provisions governing the transportation, handling, storage and protection of material, and equipment are included in Division 1, General Requirements (see Chapter 12). Establish various subparagraphs here that will deal with the specific requirements for handling, storage and delivery of materials, equipment, fixtures, and components that by their nature will require more detailed technical refinements and conditions.

1.5 JOB CONDITIONS

a. Specification subparagraphs dealing with the physical and environmental conditions under which the work is to be performed should be specified here. These include existing conditions, weather, temperature, humidity, and so on.

1.6 ALTERNATIVES

a. Alternatives are also specified under Division 1, General Requirements (see Chapter 12). Its inclusion here can be redundant and dangerous. It is suggested that alternatives be specified in Division 1 with only the following statement made here: "Alternatives affecting the work of this Section are specified in Section 01100, Alternatives."

1.7 GUARANTEE

a. Use this paragraph to set forth the requirements for guartees that exceed the one year guarantee normally included in the general conditions. In addition, establish a guarantee form (see sample at conclusion of this chapter) to ensure proper terminology and conditions rather than permitting the contractor to submit innocuous and self-serving guarantees.

## PART 2   PRODUCTS

### 2.1   MATERIALS

a. Specify the various materials to be used under a series of subparagraphs. Specify by means of reference standards, and descriptive, performance, or proprietary methods.

### 2.2   MIXES

a. Whether prepared on-site or off-site, specify the proportions of the materials listed above required to produce concrete, plaster, terrazzo, macadam, and so on.

### 2.3   FABRICATION AND MANUFACTURE

a. The component materials are specified in paragraph 2.1 above. The fabrication and manufacturing process is specified herein. For example, sheet metal, sound deadening, zinc coatings, and paint primers for metal doors would be specified in paragraph 2.1 The fabrication and assembly of the components to form a metal door would be specified here.

## PART 3   EXECUTION

### 3.1   INSPECTION

a. Inspection requires subparagraphs that formulate the criteria by which the subcontractor determines that the substrates to receive his work are sound, proper, and free of defects. These subparagraphs include condition of surfaces and inspection of structure.

### 3.2   PREPARATION

a. Preparation includes such subparagraph headings as field measurements, priming, and so on. For waterproofing or dampproofing it would include patching or grinding of surfaces to obtain a satisfactory base to receive these treatments.

### 3.3   INSTALLATION/APPLICATION/PERFORMANCE

a. This paragraph would include various subparagraphs detailing the requirements for installation details, construction and erection methods, and quality of workmanship.

3.4  FIELD QUALITY CONTROL

a. Tests and inspection procedures to determine the adequacy of the work completed and installed are specified herein. This would include tests for soil compaction, pile loading, concrete cylinder tests, erection tolerance inspections, and so on.

3.5  ADJUST AND CLEAN

a. Subparagraphs dealing with patching, adjustment, and cleaning would be used here to describe these requirements.

b. Patching would include the correction of honeycomb in concrete, defects in plaster or terrazzo, and so on.

c. Adjustment would include putting builders hardware into operating condition, balancing of a mechanical ventilation system, and so on.

d. Cleaning up in general terms is specified in Division 1, General Requirements (see Chapter 12). Cleaning of specific surfaces such as masonry, terrazzo, glass, and so on, is specified in this subparagraph.

It is important to note that the paragraph above and subparagraph headings are appropriate when the work of technical sections can be adequately and appropriately specified thereunder. Do not use these headings when they do not apply. Introduce new headings when applicable. Deviations are proper when awkwardness would result from too close an adherence to this rule.

Another feature that is recommended for the section format is an organized system for the internal numbering of the paragraphs and subparagraphs. Since computers can handle any format, the specifier can elect any alphanumeric system he chooses. A simple arrangement is as follows:

1.  MAJOR  PARAGRAPH  TITLE
    a.  Paragraph Heading
       1.  Subparagraph
          (a)  Subparagraph
             (1)  Subparagraph

(SAMPLE)
SECTION 07150
DAMPPROOFING
GUARANTEE
TO

OWNER: _____
PROJECT: _____
ARCHITECT: _____
REFERENCE:  Technical   Specification   Section   07150, Dampproofing.

TIME:  Period of Guarantee: 3 Years.
Starting Date:

GUARANTEE:  In accordance with Article____of the GENERAL CONDITIONS, the undersigned hereby guarantee the dampproofing against leaks resulting from defects of materials or workmanship; and further the undersigned agree that upon notification of such leaks within the guarantee period, to make the necessary repairs and replacements at the convenience of the Owner.

CONTRACTOR: _____ Date:_____

by:  _____   _____
(signed)                (printed)

SUBCONTRACTOR: _____ Date:_____

by:  _____   _____
(signed)                (printed)

# 6

## Types of Specifications

In general, there are two basic approaches to the writing of specifications: the method system and the results system. When the method system is employed, the specifier describes in detail the materials, workmanship, installation, and erection procedures to be used by the contractor in the conduct of his work operations in order to achieve the results expected. When the specifier instead elects to specify results, he places on the contractor the responsibility for securing the desired results by whatever methods the contractor chooses to use.

The method system can best be described as a descriptive specification; the results system is best described as a performance specification. An appropriate analogy can be made by comparing these approaches with building code standards. The specifications code sets forth specific materials and methods that are permitted under the law in the construction of a building. Under the performance code, materials and methods are left to the architect and engineer, provided that performance criteria for fire protection, structural adequacy, and sanitation are met. As a matter of fact, both the descriptive specification and the performance specification can be used together in the same project specification, each in its proper place, in order to achieve the prime objective.

## Descriptive Specifications

A descriptive specification can be defined as one that describes in detail the materials to be used and the workmanship required to fabricate, erect, and install the materials. Described in cookbook fashion are the materials, workmanship, installation, and erection procedures to be employed by the contractor. This approach is based on the wealth of information and experience that has been produced on known materials and methods.

The specifier is aware that if he specifies known bricks and mortar and proper workmanship techniques that have previously been used and put together in a specific fashion, the contractor can erect a quality masonry wall. As an example, a descriptive specification for a masonry wall would describe the materials to be used: the brick and mortar ingredients, composition of the mortar, tests of individual components, weather conditions during erection, workmanship involved in laying up the brick, type of brick bond, jointing, and, finally, the cleaning procedures. This allows all those concerned with specifications an opportunity to check each of the items specified. The supplier furnishes the brick and mortar as specified; the laboratory tests the components in accordance with specified test requirements; and the inspector checks the workmanship requirements so carefully specified. If the specifications have been accurately prepared, the masonry wall is erected accordingly, and the result the architect envisioned has been achieved through his minute description.

## Performance Specifications

Until the advent of systems building, the performance specification was used to a very limited extent. Buildings were designed utilizing unit materials that could be defined and specified by means of descriptive, proprietary, or reference specifications. Performance specifications were utilized primarily when the specifier required the contractor to match or obtain a result consistent with an existing situation. Specifying in this manner constituted a performance specification.

Other examples of performance specifications are involved with relatively simple requirements. Since end results are paramount, a performance specification can be defined as specifying end results by formulating the criteria for its accomplishment. The criteria for materials are established on the basis of physical properties of the end product. The criteria for equipment of a mechanical nature are established by operating characteristics. As an example, in a performance specification for a paint material, the end result is obtained by specifying or formulating the following criteria.

1. The painted surface shall withstand ten washings with a mild detergent.
2. The painted surface shall show no sign of alligatoring or crazing.
3. The painted surface shall be resistant to abrasion when using the Taber abrasive method.
4. The painted surface shall have an eggshell finish.

Another example of a performance specification is one for a complete installation of a heating system. The specification spells out the following performance requirements:

1. The heating plant shall be capable of providing an interior temperature of 70°F when the outside temperature is 0°F.
2. The heating system shall utilize No. 6 oil and shall be a hot-water system.
3. The heating elements shall be fin-type baseboard radiation.
4. Controls such as thermostats, aquastats, and other safety devices shall be provided to regulate heat and prevent explosion.

Since the advent of systems building using major assemblies and subassemblies, there developed a need for more sophisticated procedures to specify end results. Performance specifications encompassing these parameters are more fully explained in Chapter 7.

## Reference Specifications

The reference specification is one which makes reference to a standard that has been established for either a material, a test method, or an installation procedure. These standards similarly are predicated either on descriptive or performance criteria. Several reference standards are illustrated at the end of this chapter.

Before the advent of materials standards such as ASTM specifications, ANSI standards, or Federal specifications, materials were minutely described in the specifications so that the contractor was completely cognizant of what the specifier wanted. In many instances, these descriptive specifications for materials have been supplanted by the aforementioned standards. For example, in lieu of describing portland cement in detail, as to quality, fineness module, and other characteristics, the specifier now simply states that portland cement "meet the requirements of ASTM C-150, Type —." This method of specifying has resulted in a type of specification that can best be described as a reference specification. By making reference to a standard, the standard becomes a part of the specification to the same degree as descriptive or performance specification language is used.

The term "reference specifications" can similarly be applied to workmanship standards. Various trade associations, such as the Tile Council of America, the Gypsum Association, the Painting and Decorating Contractors of America, and others, have prepared standard workmanship specifications—for ceramic tile; furring, lathing, and plastering; painting; and so on—that can be incorporated by reference in project specifications. By so doing, the detailed descriptive workmanship clauses for these sections no longer need to be copied, but can simply be incorporated into the project specifications by means of the reference method.

It is essential that the architect and specifier be thoroughly familiar with the standards he incorporates in his specifications. Some standards cover several types and grades, and unless the type or grade is specifically stated, the choice then becomes the contractor's option and not the architect's. In addition, a particular type or grade may be more suited for a particular project so that it should be selected and specified by the architect in preference to another type or grade. Sometimes the types or grades apply to a specific climate or geographical area, which becomes automatic unless another quality is specified.

Most standard specifications have been developed by committees representing materials manufacturers, governmental authorities, testing agencies, consumers, and those having a general interest in the particular standard. In many cases, these standards are compromises; in some cases, only minimum property standards are established; in some instances, it may be necessary to augment or strengthen certain provisions of these standards. This can be done quite readily by modifying the standard. However, one must be certain when modifying a standard that the material can be manufactured or furnished under these modified standards.

All reference specifications used by an architect should be on file in his office. He needs these standards to make certain that the material or the installation procedure he specifies by means of these standards are satisfactory to him and are pertinent to the project. He needs them to check materials and test procedures submitted for his approval. If the architect elects to use a reference specification for workmanship or for a construction procedure taking place at the site, it will also be necessary for the resident project representative to have a copy of that reference specification since the detailed requirements are specified in the standard rather than in the basic specification. For example, the architect may refer to an American Concrete Institute Standard for cold weather concreting, which describes procedures for placing concrete in freezing temperatures; or to an ASTM specification for masonry mortar, which describes various materials and mixing proportions of mortar; or to an ANSI specification for setting ceramic tile, which describes installation procedures. A simple procedure to ensure that the inspector at the site has the specification reference is to include in the base specifications a provision requiring the contractor to furnish these standards at the same time he makes all his other submittals for review.

## Proprietary Specifications

A proprietary specification is one in which the specifier states outright the actual make, model, catalog number, and so on, of a product or the installation instructions of a manufacturer.

# 7

## Systems Building and
## Performance Specifications

In Chapter 6, performance specifications were described in their relation to basic materials and simple systems. With the advent of systems building, a design concept which rather than utilizing basic building materials as building blocks combines integrated assemblies and composites, early pioneers experimented with performance criteria and in turn with performance specifications.

The concept of performance specifications as it is related to systems building is still in its infancy. Systems building originated in Europe after World War II primarily to reconstruct the continent after the ravages of war and to hasten the building process. Its initial applications in the United States began with the California School Construction Systems Development (SCSD) in 1961. In turn, this development was followed by California University Residential Building Systems (URBS); Florida Schoolhouse Systems Project (SSP); Toronto Study of Educational Facilities (SEF); Building Systems Project (BSP), a joint study by the Public Buildings Service and the National Bureau of Standards; and by a host of other organizations.

The difficult part of the subject is to tell the experienced specifier how to write performance specifications for building components and systems, let alone the novice or student, inasmuch as there is yet no consensus. It should be possible, nevertheless, to outline the steps that have been taken in such a manner that the student and the specifier can better comprehend the direction and make an intelligent beginning.

Systems building and design are concerned more with subassemblies and composites rather than with individual materials and products. Our current test methods and standards for the evaluation of performance are based on individual components. Very few standards exist on the performance of subassemblies and composites. In addition, the design of a building system cuts across the design disciplines as we know them today. It requires the merging of architects and engineers to design composite units and, in turn, it requires a joint effort to evolve performance specifications.

First, new test methods have to be devised to cope with the requirements of subassemblies and assemblies. This requires the establishment of criteria for structural adequacy, fire resistivity, thermal conductivity, sound attenuation, and mechanical and electrical properties to provide for physical comfort by controlling heating, cooling, and illumination.

One approach to systems building requires that the design team establish the parameters for a project, setting forth aesthetic controls, with the specifications team establishing the performance characteristics required to meet these conditions. Obviously the specifications team will no longer be dealing with items of specific materials or products, but rather with the broader range of subassemblies and components. The process will require performance and results rather than description and methods. Individual or combined sections on materials and their installation, as currently specified in descriptive specifications, are replaced by systems performance specifications where technical sections establish the parameters of assemblies of floors, ceilings, walls, roofs, mechanical systems, and so forth, on the basis of life safety, acoustic environment, durability, and other recognized attributes.

The term "specifications team" is used here to denote that the criteria to be established to prepare performance specifications require the input of several professionals rather than the specifier alone. This is based on the fact that the requirements may cut across a number of design disciplines, and, in addition, no one individual has the broad knowledge that encompasses building science, manufacturing processes, cost control, and maintenance and operation to make the judgments required in the performance approach.

The most recent effort to develop a method for performance specifying is the work of the National Bureau of Standards which has developed performance criteria for two federal agencies. One is for the Department of Housing and Urban Development for use in Operation Breakthrough; the second is for the Public Buildings Service for office buildings. Both are based on a matrix consisting of attributes and built elements or subsystems (see Fig. 7.1). The term "attribute" can be defined as a performance requirement. The performance requirement is then developed on the basis of three major categories: requirement, criterion, and test. These three elements constitute a performance specification for a proposed subsystem.

The built element or subsystem, as shown in Fig. 7.1, is a major or minor component of the structure. Utilizing one built element, it can now be demonstrated how the attribute is used to develop a performance specification.

If a ceiling is selected as the subsystem (see *D* in Fig. 7.1) and the attribute, fire safety, is to be considered, the performance requirements can be developed as follows:

> Requirement 1: Provide fire safety
> Criterion 1: Maximum flame spread 25
> Test 1: ASTM E84.
> Requirement 2: Provide fire safety
> Criterion 2: Smoke development not to exceed 75
> Test 2: ASTM E84
> Requirement 3: Provide fire safety
> Criterion 3: Heat potential not to exceed 5000 btu/lb
> Test 3: Potential heat per *ASTM*, **61**, 1336–1347 (1961)

Obviously, each attribute listed for the ceiling subsystem must be investigated and performance requirements stated in terms of requirement, criterion, and test. The nature of the space in which the ceiling is used requires differing performance requirements to be developed. The specification team will include acoustical con-

| Built Elements | | | Attributes | | | | | | | | |
|---|---|---|---|---|---|---|---|---|---|---|---|
| | | | STRUCTURAL SERVICEABILITY | STRUCTURAL SAFETY | HEALTH AND SAFETY | FIRE SAFETY | ACOUSTIC ENVIRONMENT | ILLUMINATED ENVIRONMENT | ATMOSPHERIC ENVIRONMENT | DURABILITY/TIME RELIABILITY (FUNCTION) | SPATIAL CHARACTERISTICS AND ARRANGEMENT |
| | | | 1 | 2 | 3 | 4 | 5 | 6 | 7 | 8 | 9 |
| | STRUCTURE | A | | | | | | | | | |
| INTERIOR SPACE DIVIDERS | WALLS AND DOORS, INTER-DWELLING | B | | | | | | | | | |
| | WALLS AND DOORS, INTRA-DWELLING | C | | | | | | | | | |
| | FLOOR-CEILING | D | | | | | | | | | |
| EXTERIOR ENVELOPE | WALLS, DOORS AND WINDOWS | E | | | | | | | | | |
| | ROOF-CEILING, GROUND FLOOR | F | | | | | | | | | |
| | FIXTURES AND HARDWARE | G | | | | | | | | | |
| | PLUMBING | H | | | | | | | | | |
| | MECHANICAL EQUIPMENT, APPLIANCES | I | | | | | | | | | |
| | POWER, ELECTRICAL DISTRIBUTION, COMMUNICATIONS | J | | | | | | | | | |
| | LIGHTING ELEMENTS | K | | | | | | | | | |
| | ENCLOSED SPACES | L | | | | | | | | | |

sultants, fire safety consultants, materials experts, and others as necessary to obtain as complete input as possible.

In specifying performance, the process is much more difficult since the specifier and specification team are breaking new ground and must have the foresight to specify all parameters of a component or an assembly to assure that the requirements are properly evaluated and assessed. Even though the attributes developed by some agencies, as noted at the close of this chapter, may serve as a checklist, the team must encompass all the design disciplines in order to master the development of a performance specification. The danger, however, is that the concept is so new that some performance requirements may be overlooked.

It is suggested that anyone considering a performance specification approach to the design of building systems obtain and review the following documents to find more detailed information on attributes and performance requirements.

1. "Guide Criteria for the Evaluation of Operation Breakthrough Housing Systems," Department of Housing and Urban Development.

2. "Performance Specification for Office Buildings," Public Buildings Service.

3. "CSI Document MP-2D," Organization and Format for Performance Specifying.

4. "CSI Document MP-3F," Performance Specifications.

5. "Performance Specifications Writing for Building Components," Document D.C.9, British Ministry of Public Buildings and Works.

# 8

## Specifications Writing Techniques

### General

Specification writing techniques embody certain presentation methods of information and instructions peculiar to this literary form, and are therefore different from an essay or a novel. The specifications are written instructions intended to complement the graphic illustrations. Since both documents are combined to convey the entire message, the information contained in the specifications should be presented in a form that interlocks and does not overlap nor contradict.

### Scope of Work

A common form of duplication in specification writing that is superfluous and that can be dangerous is the use of a heading entitled "Scope of Work," or "Work Included," under which the work specified in detail in the ensuing section is summarized in outline form under this heading. Many specifiers may disagree with this assertion, and I cannot hope to make any converts out of this group. However, a review of the fundamentals of specification writing will convince the specifications trainee that the scope of work subhead, as written by some practitioners, is redundant, dangerous, time-consuming, and simply amounts to padding the specifications.

The danger in preparing a scope of work lies in duplication. The difficulties created by duplication are elaborated under the heading "Duplication—Repetition" in this chapter. For example, there have been specifications with a scope of work written for masonry which goes into such detail as follows:

The work under this contract shall include all labor and materials required for the construction of the masonry work as follows:

1. Exterior face brick in cavity wall construction with concrete block backup.
2. Exterior face brick with stone concrete backup.
3. Exterior face brick with common brick backup for parapets.
4. Common brick for interior partitions where noted.
5. Concrete block for backup in exterior masonry walls.
6. Concrete block for interior partitions where noted.
7. Structural facing tile soaps at exterior walls.

This is not quite the end of the scope of work, as it goes on ad infinitum, ad nauseum. What has the specifier accomplished? Has he given the estimator information to price the work, the builder's superintendent directions in construction, or the architect's supervisor a check on the character and quality of materials and workmanship?

The drawings, if properly rendered, indicate the location of all the materials. The specifications should not and need not describe their location since the draftsman can make subsequent changes without notifying the specifier. Another danger which sometimes results is that the scope of work list is not expanded on later in the specification, leaving only a brief outline in the scope of work that is incomplete and forms no sound basis for bidding. The estimator cannot use the scope of work as complete for fear that he will not make a comprehensive takeoff. The danger with the scope of work paragraphs is that they are not complete, but only indicate the major portions of the work under the section. The estimator can accept the scope of work as complete and fail to read the remainder of the specifications, which contains other information essential for an accurate estimate.

There may be some items listed in the scope of work that are not completely described in the specifications, whereas there are other items of work sometimes described in the specifications, but not listed in the scope. A contractor may contend that he should not be required to furnish anything not listed in the scope of work. Lawsuits have been started on lesser grounds, but this is not the only problem. It is the incident trouble and annoyance to the owner and the possible delay to the job that must be avoided. The argument in favor of the scope of work clauses is that they are a convenience to the contractor, but such clauses tend to lead the estimator, who is pressed for time, into the too common error of accepting the scope of work as sufficient in itself—with disastrous results.

Article 4.4.1 of the *AIA General Conditions* states that the contractor shall include all labor and materials necessary for the proper execution of the work. The general conditions, in turn, are part of the contract documents, and when the technical sections are written specifying clearly all materials, labor, and everything necessary to secure the construction of all that part of the building properly included in that technical section, a scope of work becomes redundant.

In general, the section title should be indicative of the scope of the section, and the table of contents is useful in alerting a contractor to any subdivision of similar work. For example, the section "Concrete Work" by itself in the table of contents would indicate that this section included all concrete work; whereas a table of contents that included such sections as "Concrete Roads and Walks," "Concrete Work," and "Precast Architectural Concrete" would inform the contractor that there is a subdivision of these items of concrete. Similarly, if the table of contents listed only "Unit Masonry," then all masonry work would be included under this heading; whereas a table of contents listing "Brickwork," "Structural Facing Tile," and "Gypsum Blockwork" would alert the contractor to a breakdown of masonry work shown on the drawings as being specified under separate section titles.

There are instances, however, where a section title may not necessarily be completely informative, and a delineation of the work included under the section may be required. For example, the section title "Curtain Wall" can be used in one specification, but its content may include glazing, sealing, venetian blind pockets, and convector enclosures. In another specification the section title "Curtain Wall" may be limited to only the metal framing and metal panels, with glazing, sealing, and other items specified under their respective sections. In this instance, a comprehensive scope of work would be appropriate to define the content of the section entitled "Curtain Wall."

If it is necessary to provide the contractor with an itemized list of the subjects contained in the specifications, it can be furnished in the form of a complete table of contents. This is quite evidently a convenience only, and an omission cannot do the legal harm that might be caused by an incomplete statement of work included under the scope of work.

In effect, when the specifier utilizes a section title for a scope of work, or if he writes an abbreviated scope of work as follows:

> "The work under this section of the specifications includes all labor, materials, equipment and services necessary to complete the concrete work as shown on the drawings and herein specified."

he has specified all concrete work under this one section. There is, then, no reason for him to enumerate concrete foundations, pits, walls, slabs, beams, and girders. It should be obvious that if the drawing indicates an item to be concrete, a specification for concrete materials and the placing of same has included all concrete shown. This simplified scope and the heading "Work of Other Sections," which is described next, should be sufficient to define what is and what is not the work covered by a specific technical section.

### Work of Other Sections

The heading "Work of Other Sections" should be reserved to exclude from a section those items a contractor might normally expect to find under a specific section, but which the specifier for good and sufficient reasons has elected to specify under another section. For example, under the earthwork section the heading "Work of Other Sections" could list the following:

1. Excavation, trenching, and backfilling for mechanical and electrical work are specified under their respective sections.

2. Furnishing of topsoil is specified under "Lawn and Planting."

Under the concrete section the heading "Work of Other Sections" could list the following:

1. Concrete bases for mechanical and electrical equipment are specified under their respective sections.

By utilizing the section title "Concrete Work" as a scope of work, or by writing an abbreviated scope of work in the manner previously illustrated, the specifier, in effect, is stating that all concrete work is specified in this section, and that the only exceptions are the concrete bases for mechanical and electrical equipment, which are listed under "Work of Other Sections." It is far simpler and safer to exclude an item by the device of the "Work of Other Sections" than to attempt to enumerate under a scope of work the sum of the parts that make up the whole.

Unfortunately, there are some specifications that use headings such as "Work by Others" and "Work Not Included" as a substitute for "Work of Other Sections." These headings can be misleading, inasmuch as they imply that the work listed under these headings are not part of the contract. The heading "Work Not Included" should be reserved for, and used only for, listing those items that are not to be included as part of a contract.

The heading "Work of Other Sections" should not list related items which are not pertinent to the scope of a particular section. For example, under the heading "Work of Other Sections," in a built-up roofing section, the following has been listed:

> *Work of Other Sections*
> 1. Membrane waterproofing
> 2. Dampproofing

Any subcontractor understands that work which is in no way related to his own is naturally not included, especially if it is not mentioned in the section. It is only work that reasonably might be inferred to be part of this work that should be listed as specified under the work of another section when that is the architect's intention.

A heading "Work Not Included," if properly used, should not be encumbered with work that is not normally done, and can be illustrated by a typical paragraph found in a painting section as follows:

*Work Not Included*
1. Painting of asphalt tile
2. Painting of glass
3. Painting of marble

Certainly, if the specifier describes paint materials and their application on specific surfaces—such as wood, ferrous metal, plaster, and concrete block—the contractor will not paint asphalt tile, glass, and marble, whether listed under the "Work Not Included" heading or not.

## Grandfather Clauses

Individuals who are not properly grounded in the principles of specification writing habitually fall back on general and all-inclusive language, which often results in what are termed "grandfather clauses" by specifiers and "murder clauses" by contractors—clauses that embrace everything, yet fail to be specific. A typical example of a grandfather clause might read as follows: "the contractor shall furnish and include everything necessary for the full and complete construction of the building whether shown or specified or not shown or described." When an architect is incompetent, he entrenches himself behind such a series of clauses, which may be interpreted to mean anything or nothing. In their failure to be specific, these clauses will, during the course of construction, require interpretations by the architect that may be difficult to enforce.

A clause such as "concrete floors shall be finished level as approved by the architect" without stating a tolerance means to the contractor, "Guess what I will make you do." An instruction to a contractor by means of a drawing or a specification must be specific, and no architect should expect a contractor to fulfill a nonspecific requirement.

## *Residuary Legatee*

Where several different kinds or classes of similar materials are used, they should be described in a manner that permits some material to be specified for every part of the building. Such a technique has been borrowed from the legal profession and is a system known as the residuary legatee. To illustrate, let us assume that in preparing a will an individual wishes to leave the bulk of his estate to his wife, but wishes to make several minor bequests to his children or to relatives. He first enumerates his minor bequests and then states, in substance, "the residue of my property I bequeath to my wife." She is then known as the "residuary legatee."

In applying this principle to specification writing, the materials occurring in the smallest quantity or in the fewest places should be listed first, and the material occurring in the remaining places becomes the residuary legatee and can be covered by some such phrase as "the rest of the building."

As example of this technique the following samples are offered:

1. In specifying glass one can list the following:
    a. Obscure glass—all toilet rooms
    b. Tempered glass—entrance doors and side lights
    c. Plate glass—borrowed lights
    d. Window glass—all other locations
2. In specifying paint:
    a. Plaster surfaces in toilets—semigloss enamel
    b. Plaster surfaces in kitchens—gloss enamel
    c. Plaster surfaces in bedrooms—flat enamel
    d. All other plaster—latex emulsion paint
3. In specifying concrete:
    a. 2500 psi concrete—concrete foundations
    b. 3000 psi concrete—concrete pavements
    c. 3500 psi concrete—all other concrete work

If this method is followed, some material will always be specified for every part of the building, whereas any other plan obliges the specifier to check all his listings most carefully for fear of not including some minor portion.

## Duplication-Repetition

In Chapter 2 it was noted that the necessary information for the construction of a building is communicated to a contractor in two forms, graphic (the drawings) and written (the specifications), and that these documents should complement one another. If this information overlaps, there can be duplication which may lead to a difference in instructions and disagreement as to which is the proper document to follow.

If this duplication were exact in each instance and remained so, it might be harmless at best; but too often the information presented on the drawings and that specified either does not agree in the first place, or, owing to last minute changes, errors and differences develop which create entirely new meanings. Repetition in the contract documents is always dangerous and should be avoided.

Technically, duplication is an exact repetition, word for word, of a sentence or a paragraph in a specification, or else it is an exact repetition of a detail on a drawing. For example, a steel ladder might be detailed on a drawing, giving the size of the side members and the diameter and spacing of the rungs. The specification should describe the quality of the material and how the rungs are let into the side members, but it should not repeat the sizes and spacing since the drawing may be altered by the draftsman, with a resulting conflict in the two documents. The unnecessary expense involved in writing and reproducing statements that merely repeat may be minor in comparison to the ultimate cost to the owner of mistakes in specification interpretation.

An exact duplication in the specification or drawing should cause no misunderstanding. However, it is seldom that we see an exact duplication. In most cases the specifier attempts to avoid duplication or repetition by stating in different words what has been said or stated elsewhere, in order to amplify. But it is precisely in attempting to amplify or reiterate in different words that conflict and ambiguity occur. It is therefore good practice to make a statement only once; if it is not satisfactory, it should be discarded and rewritten, rather than amplified or explained in other terms.

# 9

## *Bidding Requirements*

### *General*

Bidding requirements consist of documents that are used in the solicitation of bids by an owner or an agency, and are directed to bidders who might be interested in submitting bids for a project. These documents consist of three essential forms dealing with advertising, or notifying interested bidders of the existence of a proposed project; instructions pertaining to the submission of a proposal or bid; and the sample form on which the bid is to be executed by a bidder.

The three documents are Invitation to Bid, Instructions to Bidders, and Bid Form. Because of the varying practices of individual specifiers and the lack of order and terminology of the material preceding the technical specifications, some chaos and nonuniformity in the arrangement and nomenclature of these documents has existed previously.

Bidding requirements are not specifications. The basic difference lies in the fact that the former applies to a bidder prior to making an award, whereas the latter applies to a contractor and his obligations after an award is made. Generally speaking, certain information contained in the bidding requirements which is pertinent to a contractor's obligations—such as time for completion, base bid, alternatives, and unit prices—should be entered into the agreement or contract form to insure its fulfillment by the contractor.

Invitations to Bid are generally circulated in the case of private work to certain selected bidders, and in the case of public agencies, they are advertised in local newspapers. In any event, the Invitation to Bid, along with the Instructions to Bidders and a sample copy of the Bid Form, should be bound in the specifications.

### Invitation to Bid

Other terms have been used for Invitation to Bid, but are used somewhat incorrectly as the heading for this document. These include Advertisement to Bid (sometimes used in public work for public advertising), Notice to Bidders, and Notification to Contractors. The term "Invitation to Bid" is preferred since it best describes the intent of this document. The purposes of an Invitation to Bid are to attract bidders in sufficient numbers to ensure fair competition, and to notify all parties who might be interested in submitting proposals. It should be limited to information that will tell a prospective bidder whether the work is in his line, whether it is within his capacity, and whether he will have the time to prepare a bid prior to opening. It should be brief and simple, and free from extraneous and irrelevant subject matter not consistent with its purpose. It should consist of the following elements:

1. Project Title. State the name of the project, its location, and project number if any.

2. Identification of Principals. State the name and address of the architect or issuing agency, together with the date of issue.

3. Time and Place for Receipt of Bids. State the time and place where bids will be received and whether they will be publicly opened. If opened privately, indicate whether prime bidders can attend.

4. Project Description. Provide a brief but adequate description of the project, setting forth size, height, and any unusual features so that the bidder will be in a position to determine whether he has the financial and technical ability to undertake the construction of the project.

5. Type of Contract. State whether bids are being solicited for a single or segregated contract and on what basis.

6. Examination and Procurement of Documents. State where the contract documents can be examined and when and where they can be obtained. Indicate whether a deposit or a charge will be required for procurement of the documents and whether there will be any refunds.

7. Bid Security. State whether a Bid Bond or other type of bid guarantee will be required to ensure the execution of the contract.

8. Guarantee Bonds. State whether Performance Bonds and Labor and Materials Payment Bonds will be required to ensure the completion of the contract.

## Instructions to Bidders

The Instructions to Bidders have also been identified by other terms, such as Information for Bidders and Conditions of Bid. The purpose of the Instructions to Bidders is to outline the requirements necessary to prepare and submit a bid properly. As such, they are truly detailed instructions to a bidder; they guide him in soliciting information concerning discrepancies in the contract documents and provde him with all the information necessary to execute the bid form. AIA Document A701, Instructions to Bidders, has been developed by the AIA, and a sample form is included at the end of this chapter. The Instructions to Bidders consist of the following elements:

1. Form of Bid. Identify the form of bid and indicate the number of copies to be submitted.

2. Preparation of Bid. Describe which blank spaces in the Bid Form are to be filled in by the bidder, including base bids, alternatives, unit prices, and so on.

3. Submission of Bid. State how bids are to be sealed, addressed, and delivered.

4. Examination of Documents and Site. Instruct bidder to examine the contract documents and the site of the proposed project in order to familiarize himself with all aspects of the project.

5. Interpretation of Documents. State how discrepancies in contract documents discovered by bidders will be interpreted and resolved by the architect.

6. Withdrawal and Modification of Bids. State how bids can be withdrawn or modified prior to bid opening.

7. Award of Contract. Describe the procedure under which the award of the contract will be made.

8. Rejection of Bids. State the conditions under which the bids can be rejected.

9. Other Instructions to Bidders. State whether certain information relative to financial status, subcontractor, and substitutions are to be submitted with the Bid Form.

## Bid Form

The Bid Form, sometimes termed the Proposal Form or Form of Proposal, is a document prepared by the architect or issuing agency in order to assure similarity in the preparation and presentation of bids by bidders and to obtain a uniform basis of comparison. By using only the forms prepared by the issuing agency, the owner is assured that all bidders are submitting proposals on an equal basis.

The Bid Form is prepared in the form of a letter from the bidder to the owner, and contains the necessary blank spaces for the bidder to fill in contract prices as well as spaces for the required signatures and addresses.

The Bid Form consists of the following essential elements:

1. Addressee. State the name and address of the individual receiving bids.

2. Name and Address of Bidder. State the name of the organization and address of the bidder.

3. Project Identification. State the name of the project.

4. Acknowledgment. Provide an enumeration of the documents and a statement to the effect that the site has been visited and examined.

5. Bid Schedule. Set forth a bid list of all the major bid proposals.

6. Alternatives. Set forth a list of all alternative prices. A descrip-

tion of the alternatives should be set forth under Division I, General Requirements.

7. Unit Prices. Provide a list of unit prices and their description.

8. Time of Completion. Establish the time of completion or permit the bidder to insert his own time of completion.

9. Acknowledgment of Addenda. Provide spaces for acknowledgment of receipt of addenda by bidders.

10. Agreement to Accept Contract. State the conditions under which the bidder agrees to enter into a formal contract within a specified time.

11. Signature and Address of Bidder. Provide spaces to be filled in by the bidder for his signature, address, and seal where necessary.

Sample forms for the Invitation to Bid, the Instructions to Bidders, and the Bid Form are set forth as follows:

(SAMPLE)
## INVITATION TO BID

January 1, 1974

John Jones, Architect
123 Main Street
New York, New York

## LIBRARY BUILDING
### FIRST AVENUE AND MAIN STREET
### NEW YORK, NEW YORK

1. Sealed bids for the construction of the project above for the City of New York will be received by the office of John Jones, Architect, until 1:00 P.M. EST, February 1, 1974 (and then publicly opened).

2. In general, the building is a two-story structure including a basement, and is approximately 75 ft by 200 ft in size. The frame and slabs are reinforced concrete and the exterior consists of an aluminum curtain wall.

3. (Bids will be based on a single lump-sum contract.) Bids will be received for segregated contracts consisting of General Construction; Plumbing; Heating, Ventilating, and Air Conditioning; and Electrical Work.

4. Contract Documents may be examined on and after January 1,

1974 at the office of the Architect and at the following Builders' Exchanges:

a.

b.

c.

5. Contract Documents may be obtained at the office of the Architect on or after January 1, 1974 by depositing a check in the amount of $_____ per set, payable to the Architect. Deposits will be refunded to Bidders who return the documents in good condition within 10 days after the opening of bids.

6. Bid security in the form of a Bid Bond or certified check made payable to the City of New York, in an amount equal to 5% of the bid, will be required. No Bidder may withdraw his bid within 30 days after the actual date of the opening thereof.

7. Guaranty Bonds in the form of a Performance Bond and a Labor and Materials Payment Bond, in an amount equal to 100% of the bid, will be required.

(SAMPLE)
## INSTRUCTIONS TO BIDDERS

1. Bid Form

Attention is directed to the fact that these Specifications include a copy of Bid Form. This is for the information and convenience of Bidders and is not to be detached from the specifications, or filled out or executed. Separate duplicate copies of the Bid Form are to be submitted by the Bidder for that purpose as set forth below.

2. Preparation of Bid

a. Bids shall be submitted in duplicate on Bid Forms which will be furnished by the Architect.

b. Spaces are provided in the Bid Form for Base Bid and various unit and alternative prices. All such spaces shall be filled in on typewriter or in ink by the Bidder. Where both written words and numerical figures are given, the written words will apply in the event of conflict.

3. Submission of Bids

a. Bids will be received at the time and place set forth in the Invitation to Bid.

b. Envelopes containing bids shall be sealed, mailed, and addressed as follows:

> Mr. John Jones, Architect
> 123 Main Street
> New York, New York

Mark in lower left-hand corner "Bid for Construction of Library Building."

4. Examination of Site, Documents, and So On

Each Bidder shall visit the site of the proposed work and fully acquaint himself with conditions as they exist so that he may fully understand the facilities, difficulties, and restrictions attending the execution of the work under the contract. Bidders shall also thoroughly examine and be familiar with the drawings and the specifications. The failure or omission of any Bidder to receive or examine any form, instrument, or document or to visit the site and acquaint himself with conditions there existing shall in no way relieve the Bidder from any obligation with respect to his bid.

5. Interpretation of Documents

No oral interpretations will be made to any Bidder as to the meaning of the drawings and specifications. Every request for such an interpretation shall be made in writing and addressed and forwarded to the Architect. No inquiry received within five days of the date fixed for opening of bids will be given consideration. Every interpretation made to a Bidder will be in the form of an Addendum to the Contract Documents which, if issued, will be sent as promptly as is practicable to all persons to whom the Contract Documents have been issued. All such Addenda shall become part of the Contract Documents.

6. Withdrawal and Modification of Bids

a. Bids may be withdrawn by written or telegraphic request received from Bidders prior to the time fixed for opening.

b. Telelgraphic bids will not be considered, but modifications by telegraph or in writing will be considered if received prior to the hour set for opening.

7. Award of Contract

a. The contract will be awarded to the lowest responsible Bidder on the basis of low bid and accepted alternatives.

Or

b. The Owner reserves the right to award the contract on any basis he deems to his best interests.

8. Rejection of Bids

The Owner reserves the right to reject any or all bids when such rejection is in the interest of the Owner.

9. Other Instructions to Bidders

a. Bid security in the type and amount stated in the Invitation to Bid shall accompany the bid. The bid security shall be retained by the Owner if the Bidder fails to execute the contract, or fails to provide satisfactory Performance and Payment Bonds as required, within ten days after notice of award is mailed to the Bidder.

b. A financial statement on the form provided by the Architect shall accompany the bid (may be used where required).

(SAMPLE)

BID FORM

To:    John Jones, Architect
       123 Main Street
       New York, New York

From: _____ (Name of Bidder)

       _____ (Address of Bidder)

_____

For:   Construction of Library Building
       First Avenue and Main Street
       New York, New York

The Undersigned, having visited the site of proposed construction of the above noted project, and having familiarized himself with local conditions affecting the cost of the work and with all requirements of Contract Documents as prepared by Architect, and all Addenda to said Documents, hereby proposes to furnish all things as required by said Documents and Addenda thereto for the construction of said Library Building for the following amounts:

BID SCHEDULE

*Base Bid*

_____

_____ Dollars ($_____)

*Alternative*

The Undersigned will include the following Alternatives as described in Section 1 for the following amounts:

|  | *Add* | *Deduct* | *No Change* |
|---|---|---|---|
| Alternative No. 1 | \$_____ | \$_____ | _____ |
| Alternative No. 2 | \$_____ | \$_____ | _____ |

*Unit Prices*

Should additional work of the following categories be required, adjustment will be made to the Contract Sum at the following unit prices, which shall include all expenses, including overhead and profit. Should less work be required, the unit price will be 15% less than the price quoted for the additional work.

| *Description* | *Unit Price* |
|---|---|
| (a) General machine excavation, removed from site, per yd$^3$ | \$_____ |
| (b) Machine trench excavation, removed from site, per yd$^3$ | \$_____ |
| (c) Backfill due to extra excavation, per yd$^3$ | \$_____ |
| (d) Forms for concrete work, including stripping, per ft$^2$ | \$_____ |

*Time of Completion*

If awarded this contract, the undersigned will complete the work within _____ calendar days from the date of the notice to proceed.

(Note: Bidder shall insert time of completion.)

*Addendum Receipt*

Receipt of the following Addenda to the Contract Documents are acknowledged:

Addendum No. _____     Dated _____

Addendum No. _____     Dated _____

*Bid Acceptance*

If written notice of the acceptance of this Bid is mailed, telegraphed, or delivered to the Undersigned within 30 days after the date for opening of Bids or any time thereafter before this bid is withdrawn, the Undersigned will, within ten (10) days after the date of such mailing, telegraphing, or delivery of such notice, execute and deliver AIA, Contract Form No. A101, and furnish Performance and Payment Bond, in accordance with the specifications and Bid as accepted. The Undersigned hereby designates as his office to which such notice of acceptance may be mailed, telegraphed, or delivered:

Name of Bidder: _____

Bidder is:    Individual ( )   Partnership ( )   Corporation ( )
              Check one, as case may be

Residence of Bidder (if individual):_____

Date of Bid _____

If Bidder is a partnership, fill in the following blanks:

Name of Partners:

_____        _____

_____        _____

If Bidder is a Corporation, fill in the following blanks:

Organized under the laws of the State of _____.

Name and Home Address of the President _____

_____

Name and Home Address of Treasurer _____

_____

# THE AMERICAN INSTITUTE OF ARCHITECTS

*AIA DOCUMENT A701*

# Instructions to Bidders

*Use only with the latest Edition of AIA Document A201, General Conditions of the Contract for Construction*

## TABLE OF ARTICLES

"This document has been reproduced with the permission of The American Institute of Architects. Further reproduction is not authorized".

This document has been approved and endorsed by The Associated General Contractors of America.

Copyright © 1970 by The American Institute of Architects, 1735 New York Avenue, N.W., Washington, D. C. 20006. Reproduction of the material herein or substantial quotation of its provisions without permission of the AIA violates the copyright laws of the United States and will be subject to legal prosecution.

# INSTRUCTIONS TO BIDDERS

## ARTICLE 1

**DEFINITIONS**

**1.1** All definitions set forth in the General Conditions of the Contract for Construction, AIA Document A201, are applicable to these Instructions to Bidders.

**1.2** Bidding documents include the advertisement or invitation to bid, Instructions to Bidders, the bid form and the proposed Contract Documents including any Addenda issued prior to receipt of bids.

**1.3** Addenda are written or graphic instruments issued prior to the execution of the Contract which modify or interpret the bidding documents, including Drawings and Specifications, by additions, deletions, clarifications or corrections. Addenda will become part of the Contract Documents when the Construction Contract is executed.

## ARTICLE 2

**BIDDER'S REPRESENTATION**

**2.1** Each bidder by making his bid represents that he has read and understands the bidding documents.

**2.2** Each bidder by making his bid represents that he has visited the site and familiarized himself with the local conditions under which the Work is to be performed.

## ARTICLE 3

**BIDDING PROCEDURES**

**3.1** All bids must be prepared on the forms provided by the Architect and submitted in accordance with the Instructions to Bidders.

**3.2** A bid is invalid if it has not been deposited at the designated location prior to the time and date for receipt of bids indicated in the advertisement or invitation to bid, or prior to any extension thereof issued to the bidders.

**3.3** Unless otherwise provided in any supplement to these Instructions to Bidders, no bidder shall modify, withdraw or cancel his bid or any part thereof for thirty days after the time designated for the receipt of bids in the advertisement or invitation to bid.

**3.4** Prior to the receipt of bids, Addenda will be mailed or delivered to each person or firm recorded by the Architect as having received the bidding documents and will be available for inspection wherever the bidding documents are kept available for that purpose. Addenda issued after receipt of bids will be mailed or delivered only to the selected bidder.

## ARTICLE 4

**EXAMINATION OF BIDDING DOCUMENTS**

**4.1** Each bidder shall examine the bidding documents carefully and, not later than seven days prior to the date

for receipt of bids, shall make written request to the Architect for interpretation or correction of any ambiguity, inconsistency or error therein which he may discover. Any interpretation or correction will be issued as an Addendum by the Architect. Only a written interpretation or correction by Addendum shall be binding. No bidder shall rely upon any interpretation or correction given by any other method.

## ARTICLE 5

**SUBSTITUTIONS**

**5.1** Each bidder represents that his bid is based upon the materials and equipment described in the bidding documents.

**5.2** No substitution will be considered unless written request has been submitted to the Architect for approval at least ten days prior to the date for receipt of bids. Each such request shall include a complete description of the proposed substitute, the name of the material or equipment for which it is to be substituted, drawings, cuts, performance and test data and any other data or information necessary for a complete evaluation.

**5.3** If the Architect approves any proposed substitution, such approval will be set forth in an Addendum.

## ARTICLE 6

**QUALIFICATION OF BIDDERS**

**6.1** If required, a bidder shall submit to the Architect a properly executed Contractor's Qualification Statement, AIA Document A305.

## ARTICLE 7

**REJECTION OF BIDS**

**7.1** The bidder acknowledges the right of the Owner to reject any or all bids and to waive any informality or irregularity in any bid received. In addition, the bidder recognizes the right of the Owner to reject a bid if the bidder failed to furnish any required bid security, or to submit the data required by the bidding documents, or if the bid is in any way incomplete or irregular.

## ARTICLE 8

**SUBMISSION OF POST-BID INFORMATION**

**8.1** Upon request by the Architect, the selected bidder shall within seven days thereafter submit the following:

**8.1.1** A statement of costs for each major item of Work included in the bid.

**8.1.2** A designation of the Work to be performed by the bidder with his own forces.

**8.1.3** A list of names of the Subcontractors or other persons or organizations (including those who are to furnish materials or equipment fabricated to a special

AIA DOCUMENT A701 • INSTRUCTIONS TO BIDDERS • APRIL 1970 EDITION • AIA® • © 1970
THE AMERICAN INSTITUTE OF ARCHITECTS, 1735 NEW YORK AVE., N.W., WASHINGTON, D.C. 20006

design) proposed for such portions of the Work as may be designated in the bidding documents or, if no portions are so designated, the names of the Subcontractors proposed for the principal portions of the Work. The bidder will be required to establish to the satisfaction of the Architect and the Owner the reliability and responsibility of the proposed Subcontractors to furnish and perform the Work described in the Sections of the Specifications pertaining to such proposed Subcontractors' respective trades. Prior to the award of the Contract, the Architect will notify the bidder in writing if either the Owner or the Architect, after due investigation, has reasonable and substantial objection to any person or organization on such list. If the Owner or Architect has a reasonable and substantial objection to any person or organization on such list, and refuses in writing to accept such person or organization, the bidder may, at his option, withdraw his bid without forfeiture of bid security, notwithstanding anything to the contrary contained in Paragraph 3.3. If the bidder submits an acceptable substitute with an increase in his bid price to cover the difference in cost occasioned by such substitution, the Owner may, at his discretion, accept the increased bid price or he may disqualify the bidder. Subcontractors and other persons and organizations proposed by the bidder and accepted by the Owner and the Architect must be used on the Work for which they were proposed and accepted and shall not be changed except with the written approval of the Owner and the Architect.

## ARTICLE 9

**PERFORMANCE BOND AND**
**LABOR AND MATERIAL PAYMENT BOND**

**9.1** The Owner shall have the right, prior to the execution of the Contract, to require the bidder to furnish bonds covering the faithful performance of the Contract and the payment of all obligations arising thereunder in such form and amount as the Owner may prescribe and with such sureties secured through the bidder's usual sources as may be agreeable to the parties. If such bonds are stipulated in the bidding documents, the premiums shall be paid by the bidder; if required subsequent to the submission of bids, the cost shall be reimbursed by the Owner. The bidder shall deliver the required bonds to the Owner not later than the date of execution of the Contract, or if the Work is commenced prior thereto in response to a letter of intent, the bidder shall, prior to commencement of the Work, submit evidence satisfactory to the Owner that such bonds will be issued.

**9.2** The bidder shall require the attorney in fact who executes the required bonds on behalf of the surety to affix thereto a certified and current copy of his power of attorney indicating the monetary limit of such power.

AIA DOCUMENT A701 • INSTRUCTIONS TO BIDDERS • APRIL 1970 EDITION • AIA® • © 1970
THE AMERICAN INSTITUTE OF ARCHITECTS, 1735 NEW YORK AVE., N.W., WASHINGTON, D.C. 20006

3

# 10

## Bonds, Guarantees, Warranties

### Surety Bonds

Article 7.5 of the *AIA General Conditions* describes a provision for the furnishing of bonds by a contractor covering the faithful performance of the contract and for the payment of obligations arising thereunder. Standard forms have been prepared by the AIA in cooperation with the surety industries and are recommended for use in all private and public construction where a statutory form is not prescribed. These forms are AIA Document A311, Performance Bond, and Labor and Materials Payment Bond. Sample copies are illustrated at the conclusion of this chapter.

The surety bonds, sometimes called guarantee bonds or construction bonds, are an essential part of today's construction procedures. They make it possible for the contractor to provide the owner with the guarantee of a responsible surety company and that he will satisfactorily perform the project at his price and pay his bills. Of additional interest to the architect is the fact that extra architectural compensation resulting from a contractor's default caused by delinquency or insolvency is reimbursable by the use of these bonds.

The following extracts from the *AIA Architect's Handbook of Professional Practice*, Chapter 7, 1969 Edition, provide further insight into surety bonds. *These extracts have been reproduced with the permission of the American Institute of Architects. Further reproduction is not authorized.*

*Surety Bonds*

A Surety Bond is an agreement under which one party, called a Surety, agrees to answer to another party, called an Obligee, for the debt, default, or failure of a third party, called the Principal, to fulfill his contract obligations. As Surety Bonds are used in the construction industry, the Surety is usually a corporation which underwrites or guarantees construction bonds; the Obligee is usually the Owner; and the Principal ordinarily is the Contractor.

Surety Bonds, sometimes called Construction Bonds, make it possible for the Contractor to assure the Owner that the Project will be completed in accordance with the Contract Documents including payment of all obligations.

A Bond does not impose on the Surety any obligations which are separate and distinct from or additional to those assumed by the Principal in the Contract Documents. Under any Surety Bond, the Principal is primarily responsible, and every obligation of the Surety is also that of the Principal. A bond is not a substitute for the integrity, financial worth, experience, equipment, and personnel of the Contractor. Nor is such a bond an independent undertaking by the Surety so long as the Principal performs in accordance with the terms of the Contract Documents.

*Contractor Defaults*

One of the chief causes for a contractor's default is the inadequacy of his bid, either on the owner's contract or on other past, current, or future contracts. This can arise from a variety of circumstances such as deficient cost and other accounting records, unforeseen price rises, delays, labor troubles, defaults of subcontractors, through disability of the contractor or of key men, lack of adequate working capital, and tax obligations.

Bonds, in accordance with the terms of the contract, provide protection against loss resulting from the failure of others to perform. Whereas the liability of the principal for damage may be theoretically unlimited, that of the surety is limited to a certain sum of money called the "penalty" or the "penal sum," which is set out in the bond. The nature of such an instrument is an extension of credit to the principal, not in the sense of a loan of funds, but rather as an endorsement. The Performance Bond directly increases the financial responsibility of the contractor for the benefit of the owner by the amount of its penal sum.

*Relationship to Contracts*

Before the owner enters into a construction contract without a bond, it is advisable for the architect to suggest to the owner that he seek advice of legal counsel. If the owner decides not to require a bond, the architect should consider writing a letter noting such a decision.

No bond can correct the deficiencies of an inadequately drawn basic contract that may not clearly express the intentions of the parties. Even with a good basic contract, it is advisable for the architect to remind the owner to observe its terms since the contractor's surety has relied thereon in its consideration of the risk. If changes of a material nature are to be made during the course of the performance of the contract, the surety's consent to such changes is necessary to assure its continued liability under the bond.

### Amount of Bonds

A Performance Bond and a Labor and Materials Payment Bond, each in the amount of 100% of the contract price, are recommended. Where a public body is the owner, its legal counsel should obtain complete information regarding the legal requirements, amount, and form of the bond and provide this information to the architect for reference purposes.

### Statutory and Nonstatutory Bonds

Surety bonds fall into two basic categories: statutory and nonstatutory or private. Statutory bonds are those required by law. Some states have statutory provisions relating to bonds, and on private projects counsel for the owner may suggest special requirements. Therefore there is no standard form of surety bond applicable to every project. The AIA Document A311, Performance Bond, and Labor and Materials Payment Bond are steps toward such standardization, and their use is urged for all private and public contracts where a statutory form is not prescribed.

### Bond Types and Functions

• Bid Bond, AIA Document A310—The function of a Bid Bond is to guarantee that, if awarded the contract within the time stipulated, the bidder will enter into the contract and furnish the prescribed Performance Bond and Labor and Materials Payment Bond. If he fails to do so without justification, there shall be paid to the owner the difference, not to exceed the penal sum of the bond, between his bid and such larger amount for which the owner may in good faith contract with another party to perform the work covered by said bid. A Bid Bond does not assure the owner that he will get a building for the bid price. These are the functions of the Performance Bond and Labor and Materials Payment Bond.

On private work, the acceptance of a Bid Bond as an alternative to the deposit of a certified check or a bank draft is discretionary with the owner. In such cases, it is recommended that the bid security be not less than 10% of the amount of the bid, and that this sum should be expressed in words and figures as a specific number of dollars and not as a percentage of the bid. On public work, the amount of the bid security and its form may be specified by law or regulation and such legal requirements will govern.

• Performance Bond, AIA Document A311—The function of the Performance Bond is to assure the owner that the contractor will perform all the terms and conditions of the contract between him and the owner and in default thereof to protect the owner against loss up to the amount of the bond penalty. AIA Standard Forms of Agreement between the Owner and Contractor and between the Owner and Architect include provisions for payment to the architect for the added burden to him in arranging for the work to proceed should the contractor default. As an element of damage in the Agreement between Owner and Contractor, reimbursement to the owner for such additional fees is covered by this document, which also contains a Labor and Materials Payment Bond as a bond separate and distinct from the Performance Bond. The Labor and Materials Payment Bond protects laborers and material men.

Historically, the Performance Bond was the first to be developed. In the course of time, provisions for the payment of labor and material furnished in the execution of the work were added to the same form. This type of instrument is generally referred to as a Combination Performance and Payment Bond. Combination Bonds no longer carry the approval of the AIA. The two-bond system, of which AIA Document A311 is an example, is regarded as greatly superior. The inclusion in one instrument of the obligation for the performance of the contract and for the payment of laborers and material men has given rise to procedural difficulties in the handling of claims against the bond. These difficulties result from the competing interests of the owner on the one hand and laborers and material men on the other. Under the two-bond pattern, the surety is enabled to make payment without awaiting a determination as to the owner's priority. These bonds are issued by the companies as a "package" and there is no additional premium by reason of the separate Labor and Materials Payment Bond.

• Labor and Material Payment Bond, AIA Document A311—This bond guarantees that the surety will pay the contractor's bills for labor and materials in the event of the default of the contractor.

## Guarantee—Warranty

Article 4.5.1 of the *AIA General Conditions* states that "the Contractor warrants and guarantees to the Owner and the Architect that all materials and equipment incorporated in the project will be new, and that all work will be of good quality, free from faults and defects and in conformance with the Contract Documents."

Article 4.5.2 of the *AIA General Conditions* states that "the warranties and guarantees provided in this paragraph and elsewhere in the Contract Documents shall be in addition to and not in limitation of any other warranty or guarantee or remedy required by law or by the Contract Documents."

Article 9.3.3 of the *AIA General Conditions* states that "the Contractor warrants and guarantees that title to all work, materials and equipment, . . . will have passed to the Owner, . . . free and clear of all liens. . . ."

Article 13.2.2 of the *AIA General Conditions* states that "if, within one year after the date of Substantial Completion or within such longer period of time as may be prescribed by law or by the terms of any applicable special guarantee required by the Contract Documents, any of the work is found to be defective . . . the Contractor shall correct it promptly."

The references cited above to *guarantee* and *warranty* are intended to call attention to two words used in specifications that are often confused, and at no time more confusing than when the specifier looks up their meaning either in *Webster's Unabridged Dictionary* or in *Black's Law Dictionary*. When one reads all the definitions given in these sources, it may result in the very confusion warned against.

In prior issues of the *AIA General Conditions,* no references were made to warranties. The current *AIA General Conditions* includes both terms for materials, workmanship, and ownership of work free of liens. The specifier would be well advised to allow the language used in the *AIA General Conditions* to stand as written without alteration on his part, leaving to the owner and the owner's attorney the onus and responsibility of changing the language of the general conditions if they so choose.

A warranty is sometimes required in a construction contract in connection with the furnishing of certain materials, products, or equipment. By means of the warranty, the contractor or manufacturer certifies that the material, product, or equipment will perform as required. A warranty promises performance, and if the warranty is not as stated and if the material, product, or equipment should fail to perform as warranted, the contractor or manufacturer has committed a breach of contract and the owner may seek legal recourse for fulfillment of the warranty.

Many manufactured articles are advertised as guaranteed or warranteed. The term may mean much or nothing. The architect should investigate any such guarantee or warranty carefully before specifying materials or equipment so advertised. Sometimes there are "jokers" or disclaimers in the printed form.

He should also investigate the reputation and reliability of the concern giving the guarantee or warranty. When guarantees are called for, the architect should not fail to secure them before accepting the work. He cannot leave this to the surety furnishing the bond, as the surety does not undertake this duty.

The one year guarantee of work set forth in Article 13.2 may not be adequate for certain types of work. It is customary to include special guarantee clauses for such items as waterproofing, roofing, curtain walls, and wood doors, where defective workmanship or materials can result in damages that may not evidence themselves until a longer period of time has elapsed. Where additional protection is desired, the terms of Article 13.2 should be so modified under the appropriate technical section to provide for a longer guarantee period.

The following examples of guarantee clauses can be used to secure additional protection for defects in materials and workmanship:

1. *Guarantee.*   The Contractor hereby guarantees that the membrane waterproofing will be free from defective materials and workmanship for a period of three years after the date of Substantial Completion, as defined in Article 8.1.3 of the *AIA General Conditions*. Upon notification of any such defects within said guarantee period, the Contractor shall make all necessary repairs and replacements at no cost or expense to the Owner, in accordance with the requirements of Article 13.2 of the *AIA General Conditions*.

2. *Guarantee.*   Prior to final payment, the Contractor shall furnish the Architect with a three-year written guarantee of all work performed and materials furnished for membrane waterproofing. The guarantee shall state that the membrane waterproofing will be free from defective materials and workmanship for a period of three years after the date of Substantial Completion, as defined in Article 8.1.3 of the *AIA General Conditions* and that the Contractor will make all necessary repairs and replacements at no cost or expense to the Owner. The guarantee shall be in a form satisfactory to the Architect.

3. *Guarantee.*   Membrane waterproofing shall be guaranteed

against defective materials and workmanship for a period of three years after the date of Substantial Completion, as defined in Article 8.1.3 of the *AIA General Conditions*. The Contractor shall furnish the Owner with a three year guarantee bond of a Surety Company approved by the Owner, guaranteeing that the Contractor will correct and repair at his own cost and expense any defective material and workmanship in the membrane waterproofing.

Example 1 requires no additional written guarantee since the language states that "the Contractor hereby guarantees" and so on, and by his signature to the agreement the contractor has furnished the necessary written guarantee. Simply stating that the work will be guaranteed to be free from defects does not assure the owner that the contractor will remedy any defects that will appear later. The language requiring that the contractor will make good any defects is desirable to bind the contractor to his obligation. Example 2 is used in those instances where the owner or his legal counsel wants the assurance of separate written guarantees. Example 3 is used where the owner or his legal counsel wants still further assurance that a third party will undertake to answer for the performance of another.

The law regarding warranties, guarantees, and guarantors or sureties varies in different states. An owner should secure an opinion from a competent lawyer, respecting the wording of any guarantee or warranty forming a part of the specifications, since the specifications are one of the contract documents. The architect should notify the owner of this necessity. The architect may prepare the contract documents, including the agreement, but the legal form should be passed on by a lawyer who is familiar with the laws relating to building.

# THE AMERICAN INSTITUTE OF ARCHITECTS

AIA Document A310

# Bid Bond

"This document has been reproduced with the permission of The American Institute of Architects. Further reproduction is not authorized".

**KNOW ALL MEN BY THESE PRESENTS,** that we

as Principal, hereinafter called the Principal, and

a corporation duly organized under the laws of the State of
as Surety, hereinafter called the Surety, are held and firmly bound unto

as Obligee, hereinafter called the Obligee, in the sum of

Dollars ($            ),
for the payment of which sum well and truly to be made, the said Principal and the said Surety, bind ourselves, our heirs, executors, administrators, successors and assigns, jointly and severally, firmly by these presents.

**WHEREAS,** the Principal has submitted a bid for

NOW, THEREFORE, if the Obligee shall accept the bid of the Principal and the Principal shall enter into a Contract with the Obligee in accordance with the terms of such bid, and give such bond or bonds as may be specified in the bidding or Contract Documents with good and sufficient surety for the faithful performance of such Contract and for the prompt payment of labor and material furnished in the prosecution thereof, or in the event of the failure of the Principal to enter such Contract and give such bond or bonds, if the Principal shall pay to the Obligee the difference not to exceed the penalty hereof between the amount specified in said bid and such larger amount for which the Obligee may in good faith contract with another party to perform the Work covered by said bid, then this obligation shall be null and void, otherwise to remain in full force and effect.

Signed and sealed this                     day of                          19

| (Witness) | | (Principal)            (Seal) |
| --- | --- | --- |
|  | | (Title) |
| (Witness) | | (Surety)               (Seal) |
|  | | (Title) |

AIA DOCUMENT A310 • BID BOND • AIA ® • FEBRUARY 1970 ED • THE AMERICAN
INSTITUTE OF ARCHITECTS, 1735 N.Y. AVE., N.W., WASHINGTON, D. C. 20006                    1

# THE AMERICAN INSTITUTE OF ARCHITECTS

*AIA Document A311*

# Performance Bond

"This document has been reproduced with the permission of⎯⎯⎯⎯⎯
The American Institute of Architects. Further reproduction is
not authorized".

**KNOW ALL MEN BY THESE PRESENTS:** that

(Here insert full name and address or legal title of Contractor)

as Principal, hereinafter called Contractor, and,

(Here insert full name and address or legal title of Surety)

as Surety, hereinafter called Surety, are held and firmly bound unto

(Here insert full name and address or legal title of Owner)

as Obligee, hereinafter called Owner, in the amount of

Dollars ($                    ),

for the payment whereof Contractor and Surety bind themselves, their heirs, executors, administrators, successors and assigns, jointly and severally, firmly by these presents.

**WHEREAS,**

Contractor has by written agreement dated                    19    , entered into a contract with Owner for

in accordance with Drawings and Specifications prepared by

(Here insert full name and address or legal title of Architect)

which contract is by reference made a part hereof, and is hereinafter referred to as the Contract.

---

AIA DOCUMENT A311 · PERFORMANCE BOND AND LABOR AND MATERIAL PAYMENT BOND · AIA ®
FEBRUARY 1970 ED. · THE AMERICAN INSTITUTE OF ARCHITECTS, 1735 N.Y. AVE., N.W., WASHINGTON, D. C. 20006          **1**

NOW, THEREFORE, THE CONDITION OF THIS OBLIGATION is such that, if Contractor shall promptly and faithfully perform said Contract, then this obligation shall be null and void; otherwise it shall remain in full force and effect.

The Surety hereby waives notice of any alteration or extension of time made by the Owner.

Whenever Contractor shall be, and declared by Owner to be in default under the Contract, the Owner having performed Owner's obligations thereunder, the Surety may promptly remedy the default, or shall promptly

1) Complete the Contract in accordance with its terms and conditions, or

2) Obtain a bid or bids for completing the Contract in accordance with its terms and conditions, and upon determination by Surety of the lowest responsible bidder, or, if the Owner elects, upon determination by the Owner and the Surety jointly of the lowest responsible bidder, arrange for a contract between such bidder and Owner, and make available as Work progresses (even though there should be a default or a succession of

defaults under the contract or contracts of completion arranged under this paragraph) sufficient funds to pay the cost of completion less the balance of the contract price; but not exceeding, including other costs and damages for which the Surety may be liable hereunder, the amount set forth in the first paragraph hereof. The term "balance of the contract price," as used in this paragraph, shall mean the total amount payable by Owner to Contractor under the Contract and any amendments thereto, less the amount properly paid by Owner to Contractor.

Any suit under this bond must be instituted before the expiration of two (2) years from the date on which final payment under the Contract falls due.

No right of action shall accrue on this bond to or for the use of any person or corporation other than the Owner named herein or the heirs, executors, administrators or successors of the Owner.

Signed and sealed this        day of        19

_____
(Witness)

                  (Principal)        (Seal)

                  (Title)

_____
(Witness)

                  (Surety)        (Seal)

                  (Title)

AIA DOCUMENT A311 • PERFORMANCE BOND AND LABOR AND MATERIAL PAYMENT BOND • AIA ®
FEBRUARY 1970 ED. • THE AMERICAN INSTITUTE OF ARCHITECTS, 1735 N.Y. AVE., N.W., WASHINGTON, D. C. 20006      2

*AIA Document A311*

# Labor and Material Payment Bond

"This document has been reproduced
The American Institute of Architects.
not authorized".

WITH PERFORMANCE BOND IN FAVOR OF THE
PERFORMANCE OF THE CONTRACT

KNOW ALL MEN BY THESE PRESENTS: that

(Here insert full name and address or legal title of Contractor)

as Principal, hereinafter called Principal, and,

(Here insert full name and address or legal title of Surety)

as Surety, hereinafter called Surety, are held and firmly bound unto

(Here insert full name and address or legal title of Owner)

as Obligee, hereinafter called Owner, for the use and benefit of claimants as hereinbelow defined, in the

amount of
(Here insert a sum equal to at least one-half of the contract price)          Dollars ($          ),
for the payment whereof Principal and Surety bind themselves, their heirs, executors, administrators,
successors and assigns, jointly and severally, firmly by these presents.

WHEREAS,

Principal has by written agreement dated          19     entered into a contract with Owner for

in accordance with Drawings and Specifications prepared by

(Here insert full name and address or legal title of Architect)

which contract is by reference made a part hereof, and is hereinafter referred to as the Contract.

---

AIA DOCUMENT A311 • PERFORMANCE BOND AND LABOR AND MATERIAL PAYMENT BOND • AIA ®
FEBRUARY 1970 ED. • THE AMERICAN INSTITUTE OF ARCHITECTS, 1735 N.Y. AVE., N.W., WASHINGTON, D. C. 20006

3

NOW, THEREFORE, THE CONDITION OF THIS OBLIGATION is such that, if Principal shall promptly make payment to all claimants as hereinafter defined, for all labor and material used or reasonably required for use in the performance of the Contract, then this obligation shall be void; otherwise it shall remain in full force and effect, subject, however, to the following conditions:

1. A claimant is defined as one having a direct contract with the Principal or with a Subcontractor of the Principal for labor, material, or both, used or reasonably required for use in the performance of the Contract, labor and material being construed to include that part of water, gas, power, light, heat, oil, gasoline, telephone service or rental of equipment directly applicable to the Contract.

2. The above named Principal and Surety hereby jointly and severally agree with the Owner that every claimant as herein defined, who has not been paid in full before the expiration of a period of ninety (90) days after the date on which the last of such claimant's work or labor was done or performed, or materials were furnished by such claimant, may sue on this bond for the use of such claimant, prosecute the suit to final judgment for such sum or sums as may be justly due claimant, and have execution thereon. The Owner shall not be liable for the payment of any costs or expenses of any such suit.

3. No suit or action shall be commenced hereunder by any claimant:

a) Unless claimant, other than one having a direct contract with the Principal, shall have given written notice to any two of the following: the Principal, the Owner, or the Surety above named, within ninety (90) days after such claimant did or performed the last of the work or labor, or furnished the last of the materials for which said claim is made, stating with substantial

accuracy the amount claimed and the name of the party to whom the materials were furnished, or for whom the work or labor was done or performed. Such notice shall be served by mailing the same by registered mail or certified mail, postage prepaid, in an envelope addressed to the Principal, Owner or Surety, at any place where an office is regularly maintained for the trans-. action of business, or served in any manner in which legal process may be served in the state in which the aforesaid project is located, save that such service need not be made by a public officer.

b) After the expiration of one (1) year following the date on which Principal ceased Work on said Contract, it being understood, however, that if any limitation embodied in this bond is prohibited by any law controlling the construction hereof such limitation shall be deemed to be amended so as to be equal to the minimum period of limitation permitted by such law.

c) Other than in a state court of competent jurisdiction in and for the county or other political subdivision of the state in which the Project, or any part thereof, is situated, or in the United States District Court for the district in which the Project, or any part thereof, is situated, and not elsewhere.

4. The amount of this bond shall be reduced by and to the extent of any payment or payments made in good faith hereunder, inclusive of the payment by Surety of mechanics' liens which may be filed of record against said improvement, whether or not claim for the amount of such lien be presented under and against this bond.

Signed and sealed this _____ day of _____ 19 _____

_____
(Witness)

_____ (Principal)   (Seal)

_____ (Title)

_____
(Witness)

_____ (Surety)   (Seal)

_____ (Title)

AIA DOCUMENT A311 · PERFORMANCE BOND AND LABOR AND MATERIAL PAYMENT BOND · AIA ®
FEBRUARY 1970 ED. · THE AMERICAN INSTITUTE OF ARCHITECTS, 1735 N.Y. AVE., N.W., WASHINGTON, D. C. 20006

4

# 11

## General Conditions and Supplements

General conditions in essence consist of provisions that establish and pertain to the legal responsibilities and relationships between the parties involved in the work, namely the owner, the contractor, and the architect or engineer.

The American Institute of Architects as well as the Consulting Engineers Council have standard preprinted general conditions for use in private work. The Federal Government (specifically the General Services Administration), several states, and many cities and municipalities similarly have standard preprinted general conditions for use on public work.

Any architect, and especially a beginner, would be well advised to use the standard *AIA General Conditions* for a private project, or those of a governmental agency if his commission involves a public project. This admonition is based on the fact that the language of the preprinted standard has stood the test of time. There are, in addition, several advantages to using general conditions of long standing. A history of use is accompanied by precedents established in the courts. The terms are familiar to the contractor as well as to the architect. A standard document provides assurance that most major contractual-legal provisions are included.

The first edition of the *AIA General Conditions* was published in 1911. Prior to that time, a document known as the "Uniform Contract," originated in 1888, was used. This earlier document was a preprinted contract form designed for use as the contractual agreement between the owner and.the contractor. In 1911, the Uniform Contract was divided into the two parts that are familiar to us today: the Agreement (the Standard Form of Agreement between Contractor and Owner for Construction of Buildings, AIA Document No. A101), and General Conditions of the Contract for Construction (AIA Document No. A201). The latter, commonly referred to as the *AIA General Conditions,* has undergone many revisions since 1911. The current issue is the 12th Edition, dated April 1970. (See sample at the end of this chapter.)

It has also been necessary to modify the standard *AIA General Conditions* to suit the requirements for a specific project. These modifications, in the form of additions, deletions, and substitutions are called supplementary conditions. However, in the process of modification, the supplementary conditions have become larded with many articles that are, in essence, work of a general nature to be performed by the contractor in order to construct the building, and not necessarily a contractual-legal responsibility. These articles have included temporary facilities (roads, offices, fences, scaffolding, toilets, and watchmen), signs, photographs, and temporary utilities (water, heat, and electricity).

Until the advent of the *CSI Format for Construction Specifications,* there was little choice but to add these nonlegal requirements to the supplementary conditions. Now with the CSI Format, the modifications to the legal requirements of the standard *AIA General Conditions* can be taken care of in the supplementary conditions, whereas the nontechnical work requirements can be included under Division 1, General Requirements. (See Chapter 12 for the scope of sections covered under Division 1.)

The current edition of the *AIA General Conditions* reflects the radical change effected through the reduction of 44 articles to 14 articles. This has been accomplished by grouping under common article headings certain related information that was previously under separate articles. While this first major improvement is a step in the right direction, since it codifies the location of information, it nonetheless retains several articles of other than legal-contractual character that could well be placed under Division 1 of the Uniform Construction Index.

While this book is limited to the subject of technical construction specifications and cannot presume to offer expertise on legal matters, it can forewarn architects and engineers about the use of general conditions and their modifications. The most important reason for having a document to cover only the legal responsibilities and relationships of the parties is that architects are not in the business of practicing law. Under no circumstances should they draw up or modify any legal provisions unless, upon doing so, these forms are forwarded to the owner and the owner's attorney for checking and approval. To protect the architect, responsibility for these legal forms and their modifications must, in turn, be accepted in writing by the owner and his attorney. Otherwise, if the architect collects his fee and subsequently trouble develops on the project for which he has drawn the contract, he may have been guilty of practicing law illegally.

By limiting the articles in the general conditions to those having only to do with the legal responsibilities and relationships, and having the owner acknowledge his responsibility in connection with them, the architect can then develop the other nontechnical provisions in the general requirements for which he is qualified by his training and expertise. The architect is similarly cautioned with respect to the insurance provisions contained in the general conditions. In amending the article pertaining to insurance (see Article 11 of AIA Document A201 illustrated herein), the architect should consult with the owner's insurance advisors to assure that the insurance provisions adequately protect the contractor and architect as well as the owner.

It has been the practice of some architects to include the standard preprinted general conditions by reference only, thereby making it a part of the contract documents. It is recommended, however, that a physical copy of the general conditions be bound with the Project Manual since there is frequent reference to this document during the course of construction.

For the small project, the AIA Document No. A201 is considered by many architects to be both too voluminous and too comprehensive, including many articles that are not appropriate nor essential to the project. To obviate the necessity of drastically amending AIA Document No. A201 in the supplementary conditions, the AIA has developed a combination agreement and general conditions, known as the AIA Short Form for Small Construction Projects, AIA Document No. A107, a copy of which is illustrated at the end of this chapter.

For information on other standard preprinted general conditions, the reader is directed to the following associations from whom copyrighted general conditions can be purchased:

> American Association of State Highway Officials
> 917 National Press Building
> Washington, D. C.
> American Public Works Association
> 1313 East 60th Street
> Chicago, Illinois 60637
> American Society of Civil Engineers
> 345 East 47th Street
> New York, New York 10017
> Associated General Contractors of America
> 1957 E Street N.W.
> Washington, D.C. 20006
> Consulting Engineers Council
> 1155 15th Street N.W.
> Washington, D.C. 20005

To modify the *AIA General Conditions,* the architect issues a document entitled "Supplementary Conditions." Some architects accomplish this by affixing additional articles to the 14 standard articles, commencing with number 15 and continuing on through all his modifications. Others start the supplementary conditions with number 1 and continue in numerical sequence with modifications to the general conditions, even though the number bears no relationship to the article number. It is recommended that the general conditions and the supplementary conditions be made a part of the contract documents by incorporating the following sample text in the bound book of specifications, and continuing in this manner.

## GENERAL CONDITIONS

1. *General Conditions*
   *The General Conditions of the Contract for Construction,* American Institute of Architects Document No. A201, 12th edition,

April 1970, hereinafter referred to as the *AIA General Conditions*, a copy of which is bound herein, shall become a part of the Contract Documents.

2. *Supplementary Conditions*

The following Supplementary Conditions contain modifications to the *AIA General Conditions* in the form of additions, deletions, and substitutions. Where any part of the *AIA General Conditions* is so modified by the Supplementary Conditions, the unaltered provisions shall remain in effect.

a. Article 1.   Contract Documents

  1. Par. 1.1   Definitions. Add the following subparagraph:

"1.1.5   Days. The term day shall mean a calendar day of 24 hours commencing at 12:00 midnight. The term working day shall mean any calendar day except Saturdays, Sundays, and legal holidays at the site of the Project."

  2. Par. 1.2   Execution, Correlation, Intent, and Interpretations. Add the following to subparagraph 1.2.3:

"The following shall constitute the order of precedence in the event that there is a conflict between the Contract Documents:

> Agreement
> General Conditions
> Supplementary Conditions
> Specifications
> Drawings
> Note: The sequence above has been
> used by some specifiers, but it is not
> an AIA condition.

Large-scale details on drawings shall take precedence over scale measurements. Where conflicting duplications are encountered in the specifications, the most expensive material or method of construction shall take precedence."

  3. Par. 1.3.   Copies Furnished and Ownership. Modify subparagraph 1.3.1 as follows:

"A total of_____copies of the Contract Documents will be furnished, free of charge. Additional copies of the Contract Documents may be obtained by the Contractor at his expense at the cost of reproduction."

b.  Article 2.    Architect
(*Note*: Further modifications to the General Conditions may be made following the outline suggested above.)

It would be wise to consult with attorneys and insurance counselors whenever the standard general conditions are to be modified. In addition, there are several sources of information on this subject that are available as follows:

> Architects' Handbook of Professional Practice
> The American Institute of Architects
> 1735 New York Avenue, N.W.
> Washington, D.C. 20006

> Architectural and Engineering Law, Second Edition
> Bernard Tomson and Norman Coplan
> Reinhold Publishing Corporation
> 600 Summer Street
> Stamford, Conn. 06904

> The AIA Standard Contract Forms and the Law
> Parker and Adams
> Little, Brown and Company
> Boston, Massachusetts

> Architectural Practice
> Cowgill and Small
> Reinhold Publishing Corporation
> 600 Summer Street
> Stamford, Conn. 06904

> Legal Responsibilities in the Practice of Architecture
>     and Engineering
> John R. Clark, Partner
> Dechert, Price and Rhoads, Attorneys
> 3 Penn Center Plaza
> Philadelphia, Pennsylvania

### Insurance

There are many risks and liabilities in the construction of a project which are of concern to the architect. The *AIA General Conditions*,

Article 11, deals with some insurance requirements, but does not include amounts nor many other insurance coverages that may be necessary to safeguard the interests of all parties to the contract, including the architect. Obviously, the contractor has the prime responsibility, but the owner and the architect may have contingent liability, and to safeguard them, they may be added to some insurance policies as additional insured.

The *AIA General Conditions* should be modified to include the policy limits which the contractor should be required to furnish. These limits vary with the size and character of the project and with its location, and they should be consistent with the inherent risks involved. The modifications can also include other types of insurance coverage, as may be determined by the owner and his attorney and insurance advisor.

In order to ascertain that the insurance specified is in force and effect on the project, the architect requires that a certificate of insurance listing all pertinent data be furnished by the contractor's insurance company. Since each insurance company issues its own certificate, and since they differ in form, it is often difficult to understand readily what is stated in a certificate without a careful analysis of each entry.

To assure ready evaluation and analysis of insurance, the AIA has prepared a special certificate form, Document No. G705 (a sample is illustrated at the end of this chapter), which provides for the listing of policy numbers, inception, and expiration dates, and for the limits of liability under the various categories.

The AIA, through its Committee on Insurance, has developed a checklist on protection, and owner's instructions regarding insurance.

The owner should determine the types and amounts of insurance he will carry and those that the contractor will be required to carry. The architect should request written confirmation of the owner's decisions and include this information in the supplementary conditions. To guide the architect in obtaining a clear expression of the owner's decisions, a set of prototype letter texts is suggested by the AIA (see sample illustrated at the end of this chapter). These should be prepared in accordance with the needs of the particular project involved.

## Insurance Definitions

In order to familiarize the architect with some of the terms used in insurance, the following is an encyclopedia of insurance terms, as defined in the AIA *Glossary of Construction Industry Terms.* Extracts of this document have been reproduced with the permission of The American Institute of Architects. Further reproduction is not authorized.

Accident (insurance terminology): A sudden, unexpected event identifiable as to time and place. See also "Occurrence."

Bodily Injury (insurance terminology): Physical injury, sickness or disease sustained by a person. See also "Personal Injury."

Builder's Risk Insurance: A specialized form of property insurance to cover Work in the course of construction. See also "Property Insurance." (Ref: Handbook Chapter 7.)

Care, Custody and Control (insurance terminology): The term used to describe a standard exclusion in liability insurance policies. Under this exclusion, the liability insurance does not apply to damage to property in the care or custody of the insured, or to damage to property over which the insured is for any purpose exercising physical control.

Certificate of Insurance: A memorandum issued by an authorized representative of an insurance company stating the types, amounts and effective dates of insurance in force for a designated insured. (Ref: AIA Document A201 and G705.)

Completed Operations Insurance: Liability insurance coverage for injuries to persons or damage to property occurring after an operation is completed but attributed to that operation. An operation is completed (1) when all operations under the Contract have been completed or abandoned; or (2) when all operations at one project site are completed; or (3) when the portion of the Work out of which the injury or damage arises has been put to its intended use by the person or organization for whom that portion of the Work was done. Completed Operations Insurance does not apply to damage to the completed Work itself.

Comprehensive General Liability Insurance: A broad form of liability insurance covering claims for bodily injury and property damage which combines under one policy coverage for all liability exposures (except those specifically excluded) on a blanket basis and automatically covers new and unknown hazards that may develop. Comprehensive General Liability Insurance automatically includes contractual liability coverage for certain types of contracts. Products Liability, Completed Operations Liability and Broader Contractual Liability coverages are available on an optional basis. This policy may also be written to include Automobile Liability.

Contractor's Liability Insurance: Insurance purchased and maintained by the

Contractor to protect him from specified claims which may arise out of or result from his operations under the Contract, whether such operations be by himself or by any Subcontractor or by anyone directly or indirectly employed by any of them, or by anyone for whose acts any of them may be liable. (Ref: AIA Documents A201 and G705 and Handbook Chapter 7.)

Contractual Liability: Liability assumed by a party under a contract. An indemnification or "hold harmless" clause is an example of contractual liability. (Ref: AIA Documents A201 and G705.)

Employer's Liability Insurance: Insurance protection for the employer against claims by employees for damages which arise out of injuries or diseases sustained in the course of their work and which are based on common law negligence rather than on liability under workmen's compensation acts. (Ref: AIA Document G705 and Handbook Chapter 7.)

Hold Harmless: See "Indemnification." See also "Contractual Liability."

Indemnification: A contractual obligation by which one person or organization agrees to secure another against loss or damage from specified liabilities.

Liability Insurance: Insurance which protects the insured against liability on account of injury to the person or property of another. See also (1) Completed Operations Insurance; (2) Comprehensive General Liability Insurance; (3) Contractor's Liability Insurance; (4) Employer's Liability Insurance; (5) Owner's Liability Insurance; (6) Professional Liability Insurance; (7) Property Damage Insurance; (8) Public Liability Insurance; (9) Special Hazards Insurance.

Loss of Use Insurance: Insurance protecting against financial loss during the time required to repair or replace property damaged or destroyed by an insured peril. (Ref: Handbook Chapter 7.)

Negligence: Failure to exercise that degree of care which a reasonable and prudent person would exercise under the same circumstances. Legal liability for the consequences of an act or omission frequently turns on whether or not there has been negligence.

Occurrence (insurance terminology): An accident or a continuous exposure to conditions which result in injury or damage, provided the injury or damage is neither expected nor intended.

Personal Injury (insurance terminology): Injury or damage to the character or reputation of a person, as well as bodily injury. Personal injury insurance usually covers such situations as false arrest, malicious prosecution, willful detention or imprisonment, libel, slander, defamation of character, wrongful eviction, invasion of privacy or wrongful entry. See also "Bodily Injury."

Property Damage Insurance: Part of general liability insurance covering injury to or destruction of tangible property, including loss of use resulting therefrom, but usually not including property which is in the care, custody and control of the insured. See also "Care, Custody and Control." (Ref: AIA Document A201 and Handbook Chapter 7).

Property Insurance: Insurance on the Work at the site against loss or damage caused by perils of fire, lightning, extended coverage (wind, hail, explosion, except steam boiler explosion, riot, civil commotion, aircraft, land vehicles and smoke), vandalism and malicious mischief and additional perils (as otherwise provided or requested). See also "Special Hazards Insurance." (Ref: AIA Document A201 and Handbook Chapter 7.)

Public Liability Insurance: Insurance covering liability of the insured for negligent acts resulting in bodily injury, disease or death of others than employees of the insured, and/or property damage. (Ref: Handbook Chapter 7.)

Special Hazards Insurance: Additional perils insurance to be included in Property Insurance (as provided in Contract Documents or requested by Contractor or at option of Owner) such as sprinkler leakage, collapse, water damage, all physical loss, or insurance on materials and supplies at other locations and/or in transit to the site. See also "Property Insurance." (Ref: AIA Document A201 and Handbook Chapter 7.)

Steam Boiler and Machinery Insurance: Special insurance covering steam boilers, other pressure vessels and related equipment and machinery. This insurance covers damage or injury to property resulting from explosion of steam boilers which is not covered by extended coverage perils.

Workmen's Compensation Insurance: Insurance covering liability of an employer to his employees for compensation and other benefits required by workmen's compensation laws with respect to injury, sickness, disease or death arising from their employment.

XCU (insurance terminology): Letters which refer to exclusions from coverage for property damage liability arising out of (1) explosion or blasting, (2) collapse of or structural damage to any building or structure, and (3) underground damage caused by by and occurring during the use of mechanical equipment

# THE AMERICAN INSTITUTE OF ARCHITECTS

AIA Document A201

# General Conditions of the Contract for Construction

*THIS DOCUMENT HAS IMPORTANT LEGAL CONSEQUENCES; CONSULTATION WITH AN ATTORNEY IS ENCOURAGED WITH RESPECT TO ITS MODIFICATION*

"This document has been reproduced with the permission of The American Institute of Architects. Further reproduction is not authorized".

## TABLE OF ARTICLES

This document has been approved and endorsed by The Associated General Contractors of America.

Copyright 1911, 1915, 1918, 1925, 1937, 1951, 1958, 1961, 1963, 1966, 1967, © 1970 by The American Institute of Architects, 1735 New York Avenue, N.W., Washington, D. C. 20006. Reproduction of the material herein or substantial quotation of its provisions without permission of the AIA violates the copyright laws of the United States and will be subject to legal prosecution.

(THIS PAGE IS BLANK)

AIA DOCUMENT A201 • GENERAL CONDITIONS OF THE CONTRACT FOR CONSTRUCTION • TWELFTH EDITION • APRIL 1970 ED.
AIA® • © 1970 • THE AMERICAN INSTITUTE OF ARCHITECTS, 1735 NEW YORK AVENUE, N.W., WASHINGTON, D.C. 20006

# INDEX

**AIA DOCUMENT A201** • GENERAL CONDITIONS OF THE CONTRACT FOR CONSTRUCTION • TWELFTH EDITION • APRIL 1970 ED.
AIA® • © 1970 • THE AMERICAN INSTITUTE OF ARCHITECTS, 1735 NEW YORK AVENUE, N.W., WASHINGTON, D.C. 20006
       **3**

*95*

AIA DOCUMENT A201 • GENERAL CONDITIONS OF THE CONTRACT FOR CONSTRUCTION • TWELFTH EDITION • APRIL 1970 ED.
AIA® • © 1970 • THE AMERICAN INSTITUTE OF ARCHITECTS, 1735 NEW YORK AVENUE, N.W., WASHINGTON, D.C. 20006

## ARTICLE 1

## CONTRACT DOCUMENTS

### 1.1 DEFINITIONS

#### 1.1.1 THE CONTRACT DOCUMENTS

The Contract Documents consist of the Agreement, the Conditions of the Contract (General, Supplementary and other Conditions), the Drawings, the Specifications, all Addenda issued prior to execution of the Contract, and all Modifications thereto. A Modification is (1) a written amendment to the Contract signed by both parties, (2) a Change Order, (3) a written interpretation issued by the Architect pursuant to Subparagraph 1.2.5, or (4) a written order for a minor change in the Work issued by the Architect pursuant to Paragraph 12.3. A Modification may be made only after execution of the Contract.

#### 1.1.2 THE CONTRACT

The Contract Documents form the Contract. The Contract represents the entire and integrated agreement between the parties hereto and supersedes all prior negotiations, representations, or agreements, either written or oral, including the bidding documents. The Contract may be amended or modified only by a Modification as defined in Subparagraph 1.1.1.

#### 1.1.3 THE WORK

The term Work includes all labor necessary to produce the construction required by the Contract Documents, and all materials and equipment incorporated or to be incorporated in such construction.

#### 1.1.4 THE PROJECT

The Project is the total construction designed by the Architect of which the Work performed under the Contract Documents may be the whole or a part.

### 1.2 EXECUTION, CORRELATION, INTENT AND INTERPRETATIONS

#### 1.2.1 The Contract Documents shall be signed in not less than triplicate by the Owner and Contractor. If either the Owner or the Contractor or both do not sign the Conditions of the Contract, Drawings, Specifications, or any of the other Contract Documents, the Architect shall identify them.

#### 1.2.2 By executing the Contract, the Contractor represents that he has visited the site, familiarized himself with the local conditions under which the Work is to be performed, and correlated his observations with the requirements of the Contract Documents.

#### 1.2.3 The Contract Documents are complementary, and what is required by any one shall be as binding as if required by all. The intention of the Documents is to include all labor, materials, equipment and other items

as provided in Subparagraph 4.4.1 necessary for the proper execution and completion of the Work. It is not intended that Work not covered under any heading, section, branch, class or trade of the Specifications shall be supplied unless it is required elsewhere in the Contract Documents or is reasonably inferable therefrom as being necessary to produce the intended results. Words which have well-known technical or trade meanings are used herein in accordance with such recognized meanings.

#### 1.2.4 The organization of the Specifications into divisions, sections and articles, and the arrangement of Drawings shall not control the Contractor in dividing the Work among Subcontractors or in establishing the extent of Work to be performed by any trade.

#### 1.2.5 Written interpretations necessary for the proper execution or progress of the Work, in the form of drawings or otherwise, will be issued with reasonable promptness by the Architect and in accordance with any schedule agreed upon. Either party to the Contract may make written request to the Architect for such interpretations. Such interpretations shall be consistent with and reasonably inferable from the Contract Documents, and may be effected by Field Order.

### 1.3 COPIES FURNISHED AND OWNERSHIP

#### 1.3.1 Unless otherwise provided in the Contract Documents, the Contractor will be furnished, free of charge, all copies of Drawings and Specifications reasonably necessary for the execution of the Work.

#### 1.3.2 All Drawings, Specifications and copies thereof furnished by the Architect are and shall remain his property. They are not to be used on any other project, and, with the exception of one contract set for each party to the Contract, are to be returned to the Architect on request at the completion of the Work

## ARTICLE 2

## ARCHITECT

### 2.1 DEFINITION

#### 2.1.1 The Architect is the person or organization licensed to practice architecture and identified as such in the Agreement and is referred to throughout the Contract Documents as if singular in number and masculine in gender. The term Architect means the Architect or his authorized representative.

#### 2.1.2 Nothing contained in the Contract Documents shall create any contractual relationship between the Architect and the Contractor.

### 2.2 ADMINISTRATION OF THE CONTRACT

#### 2.2.1 The Architect will provide general Administration of the Construction Contract, including performance of the functions hereinafter described.

AIA DOCUMENT A201 • GENERAL CONDITIONS OF THE CONTRACT FOR CONSTRUCTION • TWELFTH EDITION • APRIL 1970 ED.
AIA® • © 1970 • THE AMERICAN INSTITUTE OF ARCHITECTS, 1735 NEW YORK AVENUE, N.W., WASHINGTON, D.C. 20006          5

**2.2.2** The Architect will be the Owner's representative during construction and until final payment. The Architect will have authority to act on behalf of the Owner to the extent provided in the Contract Documents, unless otherwise modified by written instrument which will be shown to the Contractor. The Architect will advise and consult with the Owner, and all of the Owner's instructions to the Contractor shall be issued through the Architect.

**2.2.3** The Architect shall at all times have access to the Work wherever it is in preparation and progress. The Contractor shall provide facilities for such access so the Architect may perform his functions under the Contract Documents.

**2.2.4** The Architect will make periodic visits to the site to familiarize himself generally with the progress and quality of the Work and to determine in general if the Work is proceeding in accordance with the Contract Documents. On the basis of his on-site observations as an architect, he will keep the Owner informed of the progress of the Work, and will endeavor to guard the Owner against defects and deficiencies in the Work of the Contractor. The Architect will not be required to make exhaustive or continuous on-site inspections to check the quality or quantity of the Work. The Architect will not be responsible for construction means, methods, techniques, sequences or procedures, or for safety precautions and programs in connection with the Work, and he will not be responsible for the Contractor's failure to carry out the Work in accordance with the Contract Documents.

**2.2.5** Based on such observations and the Contractor's Applications for Payment, the Architect will determine the amounts owing to the Contractor and will issue Certificates for Payment in such amounts, as provided in Paragraph 9.4.

**2.2.6** The Architect will be, in the first instance, the interpreter of the requirements of the Contract Documents and the judge of the performance thereunder by both the Owner and Contractor. The Architect will, within a reasonable time, render such interpretations as he may deem necessary for the proper execution or progress of the Work.

**2.2.7** Claims, disputes and other matters in question between the Contractor and the Owner relating to the execution or progress of the Work or the interpretation of the Contract Documents shall be referred initially to the Architect for decision which he will render in writing within a reasonable time.

**2.2.8** All interpretations and decisions of the Architect shall be consistent with the intent of the Contract Documents. In his capacity as interpreter and judge, he will exercise his best efforts to insure faithful performance by both the Owner and the Contractor and will not show partiality to either.

**2.2.9** The Architect's decisions in matters relating to artistic effect will be final if consistent with the intent of the Contract Documents.

**2.2.10** Any claim, dispute or other matter that has been referred to the Architect, except those relating to artistic effect as provided in Subparagraph 2.2.9 and except any

which have been waived by the making or acceptance of final payment as provided in Subparagraphs 9.7.5 and 9.7.6, shall be subject to arbitration upon the written demand of either party. However, no demand for arbitration of any such claim, dispute or other matter may be made until the earlier of:

**2.2.10.1** The date on which the Architect has rendered his written decision, or

　　**.2** the tenth day after the parties have presented their evidence to the Architect or have been given a reasonable opportunity to do so, if the Architect has not rendered his written decision by that date.

**2.2.11** If a decision of the Architect is made in writing and states that it is final but subject to appeal, no demand for arbitration of a claim, dispute or other matter covered by such decision may be made later than thirty days after the date on which the party making the demand received the decision. The failure to demand arbitration within said thirty days' period will result in the Architect's decision becoming final and binding upon the Owner and the Contractor. If the Architect renders a decision after arbitration proceedings have been initiated, such decision may be entered as evidence but will not supersede any arbitration proceedings unless the decision is acceptable to the parties concerned.

**2.2.12** The Architect will have authority to reject Work which does not conform to the Contract Documents. Whenever, in his reasonable opinion, he considers it necessary or advisable to insure the proper implementation of the intent of the Contract Documents, he will have authority to require special inspection or testing of the Work in accordance with Subparagraph 7.8.2 whether or not such Work be then fabricated, installed or completed. However, neither the Architect's authority to act under this Subparagraph 2.2.12, nor any decision made by him in good faith either to exercise or not to exercise such authority, shall give rise to any duty or responsibility of the Architect to the Contractor, any Subcontractor, any of their agents or employees, or any other person performing any of the Work.

**2.2.13** The Architect will review Shop Drawings and Samples as provided in Subparagraphs 4.13.1 through 4.13.8 inclusive.

**2.2.14** The Architect will prepare Change Orders in accordance with Article 12, and will have authority to order minor changes in the Work as provided in Subparagraph 12.3.1.

**2.2.15** The Architect will conduct inspections to determine the dates of Substantial Completion and final completion, will receive and review written guarantees and related documents required by the Contract and assembled by the Contractor and will issue a final Certificate for Payment.

**2.2.16** If the Owner and Architect agree, the Architect will provide one or more Full-Time Project Representatives to assist the Architect in carrying out his responsibilities at the site. The duties, responsibilities and limitations of authority of any such Project Representative shall be as set forth in an exhibit to be incorporated in the Contract Documents.

**AIA DOCUMENT A201** • GENERAL CONDITIONS OF THE CONTRACT FOR CONSTRUCTION • TWELFTH EDITION • APRIL 1970 ED.
AIA® • © 1970 • THE AMERICAN INSTITUTE OF ARCHITECTS, 1735 NEW YORK AVENUE, N.W., WASHINGTON, D.C. 20006

**2.2.17** The duties, responsibilities and limitations of authority of the Architect as the Owner's representative during construction as set forth in Articles 1 through 14 inclusive of these General Conditions will not be modified or extended without written consent of the Owner, the Contractor and the Architect.

**2.2.18** The Architect will not be responsible for the acts or omissions of the Contractor, any Subcontractors, or any of their agents or employees, or any other persons performing any of the Work.

**2.2.19** In case of the termination of the employment of the Architect, the Owner shall appoint an architect against whom the Contractor makes no reasonable objection, whose status under the Contract Documents shall be that of the former architect. Any dispute in connection with such appointment shall be subject to arbitration.

## ARTICLE 3

## OWNER

**3.1**  **DEFINITION**

**3.1.1** The Owner is the person or organization identified as such in the Agreement and is referred to throughout the Contract Documents as if singular in number and masculine in gender. The term Owner means the Owner or his authorized representative.

**3.2**  **INFORMATION AND SERVICES REQUIRED OF THE OWNER**

**3.2.1** The Owner shall furnish all surveys describing the physical characteristics, legal limits and utility locations for the site of the Project.

**3.2.2** The Owner shall secure and pay for easements for permanent structures or permanent changes in existing facilities.

**3.2.3** Information or services under the Owner's control shall be furnished by the Owner with reasonable promptness to avoid delay in the orderly progress of the Work.

**3.2.4** The Owner shall issue all instructions to the Contractor through the Architect.

**3.2.5** The foregoing are in addition to other duties and responsibilities of the Owner enumerated herein and especially those in respect to Payment and Insurance in Articles 9 and 11 respectively.

**3.3**  **OWNER'S RIGHT TO STOP THE WORK**

**3.3.1** If the Contractor fails to correct defective Work or persistently fails to supply materials or equipment in accordance with the Contract Documents, the Owner may order the Contractor to stop the Work, or any portion thereof, until the cause for such order has been eliminated.

**3.4**  **OWNER'S RIGHT TO CARRY OUT THE WORK**

**3.4.1** If the Contractor defaults or neglects to carry out the Work in accordance with the Contract Documents or fails to perform any provision of the Contract, the Owner may, after seven days' written notice to the Contractor and without prejudice to any other remedy he

may have, make good such deficiencies. In such case an appropriate Change Order shall be issued deducting from the payments then or thereafter due the Contractor the cost of correcting such deficiencies, including the cost of the Architect's additional services made necessary by such default, neglect or failure. The Architect must approve both such action and the amount charged to the Contractor. If the payments then or thereafter due the Contractor are not sufficient to cover such amount, the Contractor shall pay the difference to the Owner.

## ARTICLE 4

## CONTRACTOR

**4.1**  **DEFINITION**

**4.1.1** The Contractor is the person or organization identified as such in the Agreement and is referred to throughout the Contract Documents as if singular in number and masculine in gender. The term Contractor means the Contractor or his authorized representative.

**4.2**  **REVIEW OF CONTRACT DOCUMENTS**

**4.2.1** The Contractor shall carefully study and compare the Contract Documents and shall at once report to the Architect any error, inconsistency or omission he may discover. The Contractor shall not be liable to the Owner or the Architect for any damage resulting from any such errors, inconsistencies or omissions in the Contract Documents. The Contractor shall do no Work without Drawings, Specifications or Modifications.

**4.3**  **SUPERVISION AND CONSTRUCTION PROCEDURES**

**4.3.1** The Contractor shall supervise and direct the Work, using his best skill and attention. He shall be solely responsible for all construction means, methods, techniques, sequences and procedures and for coordinating all portions of the Work under the Contract.

**4.4**  **LABOR AND MATERIALS**

**4.4.1** Unless otherwise specifically noted, the Contractor shall provide and pay for all labor, materials, equipment, tools, construction equipment and machinery, water, heat, utilities, transportation, and other facilities and services necessary for the proper execution and completion of the Work.

**4.4.2** The Contractor shall at all times enforce strict discipline and good order among his employees and shall not employ on the Work any unfit person or anyone not skilled in the task assigned to him.

**4.5**  **WARRANTY**

**4.5.1** The Contractor warrants to the Owner and the Architect that all materials and equipment furnished under this Contract will be new unless otherwise specified, and that all Work will be of good quality, free from faults and defects and in conformance with the Contract Documents. All Work not so conforming to these standards may be considered defective. If required by the Architect, the Contractor shall furnish satisfactory evidence as to the kind and quality of materials and equipment.

**4.6**  **TAXES**

**4.6.1** The Contractor shall pay all sales, consumer, use and other similar taxes required by law.

### 4.7 PERMITS, FEES AND NOTICES

**4.7.1** The Contractor shall secure and pay for all permits, governmental fees and licenses necessary for the proper execution and completion of the Work, which are applicable at the time the bids are received. It is not the responsibility of the Contractor to make certain that the Drawings and Specifications are in accordance with applicable laws, statutes, building codes and regulations.

**4.7.2** The Contractor shall give all notices and comply with all laws, ordinances, rules, regulations and orders of any public authority bearing on the performance of the Work. If the Contractor observes that any of the Contract Documents are at variance therewith in any respect, he shall promptly notify the Architect in writing, and any necessary changes shall be adjusted by appropriate Modification. If the Contractor performs any Work knowing it to be contrary to such laws, ordinances, rules and regulations, and without such notice to the Architect, he shall assume full responsibility therefor and shall bear all costs attributable thereto.

### 4.8 CASH ALLOWANCES

**4.8.1** The Contractor shall include in the Contract Sum all allowances stated in the Contract Documents. These allowances shall cover the net cost of the materials and equipment delivered and unloaded at the site, and all applicable taxes. The Contractor's handling costs on the site, labor, installation costs, overhead, profit and other expenses contemplated for the original allowance shall be included in the Contract Sum and not in the allowance. The Contractor shall cause the Work covered by these allowances to be performed for such amounts and by such persons as the Architect may direct, but he will not be required to employ persons against whom he makes a reasonable objection. If the cost, when determined, is more than or less than the allowance, the Contract Sum shall be adjusted accordingly by Change Order which will include additional handling costs on the site, labor, installation costs, overhead, profit and other expenses resulting to the Contractor from any increase over the original allowance.

### 4.9 SUPERINTENDENT

**4.9.1** The Contractor shall employ a competent superintendent and necessary assistants who shall be in attendance at the Project site during the progress of the Work. The superintendent shall be satisfactory to the Architect, and shall not be changed except with the consent of the Architect, unless the superintendent proves to be unsatisfactory to the Contractor and ceases to be in his employ. The superintendent shall represent the Contractor and all communications given to the superintendent shall be as binding as if given to the Contractor. Important communications will be confirmed in writing. Other communications will be so confirmed on written request in each case.

### 4.10 RESPONSIBILITY FOR THOSE PERFORMING THE WORK

**4.10.1** The Contractor shall be responsible to the Owner for the acts and omissions of all his employees and all Subcontractors, their agents and employees, and all other persons performing any of the Work under a contract with the Contractor.

### 4.11 PROGRESS SCHEDULE

**4.11.1** The Contractor, immediately after being awarded the Contract, shall prepare and submit for the Architect's approval an estimated progress schedule for the Work. The progress schedule shall be related to the entire Project to the extent required by the Contract Documents. This schedule shall indicate the dates for the starting and completion of the various stages of construction and shall be revised as required by the conditions of the Work, subject to the Architect's approval.

### 4.12 DRAWINGS AND SPECIFICATIONS AT THE SITE

**4.12.1** The Contractor shall maintain at the site for the Owner one copy of all Drawings, Specifications, Addenda, approved Shop Drawings, Change Orders and other Modifications, in good order and marked to record all changes made during construction. These shall be available to the Architect. The Drawings, marked to record all changes made during construction, shall be delivered to him for the Owner upon completion of the Work.

### 4.13 SHOP DRAWINGS AND SAMPLES

**4.13.1** Shop Drawings are drawings, diagrams, illustrations, schedules, performance charts, brochures and other data which are prepared by the Contractor or any Subcontractor, manufacturer, supplier or distributor, and which illustrate some portion of the Work.

**4.13.2** Samples are physical examples furnished by the Contractor to illustrate materials, equipment or workmanship, and to establish standards by which the Work will be judged.

**4.13.3** The Contractor shall review, stamp with his approval and submit, with reasonable promptness and in orderly sequence so as to cause no delay in the Work or in the work of any other contractor, all Shop Drawings and Samples required by the Contract Documents or subsequently by the Architect as covered by Modifications. Shop Drawings and Samples shall be properly identified as specified, or as the Architect may require. At the time of submission the Contractor shall inform the Architect in writing of any deviation in the Shop Drawings or Samples from the requirements of the Contract Documents.

**4.13.4** By approving and submitting Shop Drawings and Samples, the Contractor thereby represents that he has determined and verified all field measurements, field construction criteria, materials, catalog numbers and similar data, or will do so, and that he has checked and coordinated each Shop Drawing and Sample with the requirements of the Work and of the Contract Documents.

**4.13.5** The Architect will review and approve Shop Drawings and Samples with reasonable promptness so as to cause no delay, but only for conformance with the design concept of the Project and with the information given in the Contract Documents. The Architect's approval of a separate item shall not indicate approval of an assembly in which the item functions.

**4.13.6** The Contractor shall make any corrections required by the Architect and shall resubmit the required number of corrected copies of Shop Drawings or new Samples until approved. The Contractor shall direct spe-

AIA DOCUMENT A201 • GENERAL CONDITIONS OF THE CONTRACT FOR CONSTRUCTION • TWELFTH EDITION • APRIL 1970 ED.
AIA® • © 1970 • THE AMERICAN INSTITUTE OF ARCHITECTS, 1735 NEW YORK AVENUE, N.W., WASHINGTON, D.C. 20006

cific attention in writing or on resubmitted Shop Drawings to revisions other than the corrections requested by the Architect on previous submissions.

**4.13.7** The Architect's approval of Shop Drawings or Samples shall not relieve the Contractor of responsibility for any deviation from the requirements of the Contract Documents unless the Contractor has informed the Architect in writing of such deviation at the time of submission and the Architect has given written approval to the specific deviation, nor shall the Architect's approval relieve the Contractor from responsibility for errors or omissions in the Shop Drawings or Samples.

**4.13.8** No portion of the Work requiring a Shop Drawing or Sample submission shall be commenced until the submission has been approved by the Architect. All such portions of the Work shall be in accordance with approved Shop Drawings and Samples.

**4.14 USE OF SITE**

**4.14.1** The Contractor shall confine operations at the site to areas permitted by law, ordinances, permits and the Contract Documents and shall not unreasonably encumber the site with any materials or equipment.

**4.15 CUTTING AND PATCHING OF WORK**

**4.15.1** The Contractor shall do all cutting, fitting or patching of his Work that may be required to make its several parts fit together properly, and shall not endanger any Work by cutting, excavating or otherwise altering the Work or any part of it.

**4.16 CLEANING UP**

**4.16.1** The Contractor at all times shall keep the premises free from accumulation of waste materials or rubbish caused by his operations. At the completion of the Work he shall remove all his waste materials and rubbish from and about the Project as well as all his tools, construction equipment, machinery and surplus materials, and shall clean all glass surfaces and leave the Work "broom-clean" or its equivalent, except as otherwise specified.

**4.16.2** If the Contractor fails to clean up, the Owner may do so and the cost thereof shall be charged to the Contractor as provided in Paragraph 3.4.

**4.17 COMMUNICATIONS**

**4.17.1** The Contractor shall forward all communications to the Owner through the Architect.

**4.18 INDEMNIFICATION**

**4.18.1** The Contractor shall indemnify and hold harmless the Owner and the Architect and their agents and employees from and against all claims, damages, losses and expenses including attorneys' fees arising out of or resulting from the performance of the Work, provided that any such claim, damage, loss or expense (1) is attributable to bodily injury, sickness, disease or death, or to injury to or destruction of tangible property (other than the Work itself) including the loss of use resulting therefrom, and (2) is caused in whole or in part by any negligent act or omission of the Contractor, any Subcontractor, anyone directly or indirectly employed by any of them or anyone for whose acts any of them may be liable,

regardless of whether or not it is caused in part by a party indemnified hereunder.

**4.18.2** In any and all claims against the Owner or the Architect or any of their agents or employees by any employee of the Contractor, any Subcontractor, anyone directly or indirectly employed by any of them or anyone for whose acts any of them may be liable, the indemnification obligation under this Paragraph 4.18 shall not be limited in any way by any limitation on the amount or type of damages, compensation or benefits payable by or for the Contractor or any Subcontractor under workmen's compensation acts, disability benefit acts or other employee benefit acts.

**4.18.3** The obligations of the Contractor under this Paragraph 4.18 shall not extend to the liability of the Architect, his agents or employees arising out of (1) the preparation or approval of maps, drawings, opinions, reports, surveys, Change Orders, designs or specifications, or (2) the giving of or the failure to give directions or instructions by the Architect, his agents or employees provided such giving or failure to give is the primary cause of the injury or damage.

# ARTICLE 5

## SUBCONTRACTORS

**5.1 DEFINITION**

**5.1.1** A Subcontractor is a person or organization who has a direct contract with the Contractor to perform any of the Work at the site. The term Subcontractor is referred to throughout the Contract Documents as if singular in number and masculine in gender and means a Subcontractor or his authorized representative.

**5.1.2** A Sub-subcontractor is a person or organization who has a direct or indirect contract with a Subcontractor to perform any of the Work at the site. The term Sub-subcontractor is referred to throughout the Contract Documents as if singular in number and masculine in gender and means a Sub-subcontractor or an authorized representative thereof.

**5.1.3** Nothing contained in the Contract Documents shall create any contractual relation between the Owner or the Architect and any Subcontractor or Sub-subcontractor.

**5.2 AWARD OF SUBCONTRACTS AND OTHER CONTRACTS FOR PORTIONS OF THE WORK**

**5.2.1** Unless otherwise specified in the Contract Documents or in the Instructions to Bidders, the Contractor, as soon as practicable after the award of the Contract, shall furnish to the Architect in writing for acceptance by the Owner and the Architect a list of the names of the Subcontractors proposed for the principal portions of the Work. The Architect shall promptly notify the Contractor in writing if either the Owner or the Architect, after due investigation, has reasonable objection to any Subcontractor on such list and does not accept him. Failure of the Owner or Architect to make objection promptly to any Subcontractor on the list shall constitute acceptance of such Subcontractor.

AIA DOCUMENT A201 • GENERAL CONDITIONS OF THE CONTRACT FOR CONSTRUCTION • TWELFTH EDITION • APRIL 1970 ED.
AIA® • © 1970 • THE AMERICAN INSTITUTE OF ARCHITECTS, 1735 NEW YORK AVENUE, N.W., WASHINGTON, D.C. 20006

9

**5.2.2** The Contractor shall not contract with any Sub-contractor or any person or organization (including those who are to furnish materials or equipment fabricated to a special design) proposed for portions of the Work designated in the Contract Documents or in the Instructions to Bidders or, if none is so designated, with any Subcontractor proposed for the principal portions of the Work who has been rejected by the Owner and the Architect. The Contractor will not be required to contract with any Subcontractor or person or organization against whom he has a reasonable objection.

**5.2.3** If the Owner or Architect refuses to accept any Subcontractor or person or organization on a list submitted by the Contractor in response to the requirements of the Contract Documents or the Instructions to Bidders, the Contractor shall submit an acceptable substitute and the Contract Sum shall be increased or decreased by the difference in cost occasioned by such substitution and an appropriate Change Order shall be issued; however, no increase in the Contract Sum shall be allowed for any such substitution unless the Contractor has acted promptly and responsively in submitting for acceptance any list or lists of names as required by the Contract Documents or the Instructions to Bidders.

**5.2.4** If the Owner or the Architect requires a change of any proposed Subcontractor or person or organization previously accepted by them, the Contract Sum shall be increased or decreased by the difference in cost occasioned by such change and an appropriate Change Order shall be issued.

**5.2.5** The Contractor shall not make any substitution for any Subcontractor or person or organization who has been accepted by the Owner and the Architect, unless the substitution is acceptable to the Owner and the Architect.

**5.3  SUBCONTRACTUAL RELATIONS**

**5.3.1** All work performed for the Contractor by a Subcontractor shall be pursuant to an appropriate agreement between the Contractor and the Subcontractor (and where appropriate between Subcontractors and Sub-subcontractors) which shall contain provisions that:

  .1 preserve and protect the rights of the Owner and the Architect under the Contract with respect to the Work to be performed under the subcontract so that the subcontracting thereof will not prejudice such rights;

  .2 require that such Work be performed in accordance with the requirements of the Contract Documents;

  .3 require submission to the Contractor of applications for payment under each subcontract to which the Contractor is a party, in reasonable time to enable the Contractor to apply for payment in accordance with Article 9;

  .4 require that all claims for additional costs, extensions of time, damages for delays or otherwise with respect to subcontracted portions of the Work shall be submitted to the Contractor (via any Subcontractor or Sub-subcontractor where appropriate) in sufficient time so that the Con-

tractor may comply in the manner provided in the Contract Documents for like claims by the Contractor upon the Owner;

  .5 waive all rights the contracting parties may have against one another for damages caused by fire or other perils covered by the property insurance described in Paragraph 11.3, except such rights as they may have to the proceeds of such insurance held by the Owner as trustee under Paragraph 11.3; and

  .6 obligate each Subcontractor specifically to consent to the provisions of this Paragraph 5.3.

**5.4  PAYMENTS TO SUBCONTRACTORS**

**5.4.1** The Contractor shall pay each Subcontractor, upon receipt of payment from the Owner, an amount equal to the percentage of completion allowed to the Contractor on account of such Subcontractor's Work, less the percentage retained from payments to the Contractor. The Contractor shall also require each Subcontractor to make similar payments to his subcontractors.

**5.4.2** If the Architect fails to issue a Certificate for Payment for any cause which is the fault of the Contractor and not the fault of a particular Subcontractor, the Contractor shall pay that Subcontractor on demand, made at any time after the Certificate for Payment should otherwise have been issued, for his Work to the extent completed, less the retained percentage.

**5.4.3** The Contractor shall pay each Subcontractor a just share of any insurance moneys received by the Contractor under Article 11, and he shall require each Subcontractor to make similar payments to his subcontractors.

**5.4.4** The Architect may, on request and at his discretion, furnish to any Subcontractor, if practicable, information regarding percentages of completion certified to the Contractor on account of Work done by such Subcontractors.

**5.4.5** Neither the Owner nor the Architect shall have any obligation to pay or to see to the payment of any moneys to any Subcontractor except as may otherwise be required by law.

## ARTICLE 6

### SEPARATE CONTRACTS

**6.1  OWNER'S RIGHT TO AWARD SEPARATE CONTRACTS**

**6.1.1** The Owner reserves the right to award other contracts in connection with other portions of the Project under these or similar Conditions of the Contract.

**6.1.2** When separate contracts are awarded for different portions of the Project, "the Contractor" in the contract documents in each case shall be the contractor who signs each separate contract.

**6.2  MUTUAL RESPONSIBILITY OF CONTRACTORS**

**6.2.1** The Contractor shall afford other contractors reasonable opportunity for the introduction and storage of their materials and equipment and the execution of their

AIA DOCUMENT A201 • GENERAL CONDITIONS OF THE CONTRACT FOR CONSTRUCTION • TWELFTH EDITION • APRIL 1970 ED.
AIA® • © 1970 • THE AMERICAN INSTITUTE OF ARCHITECTS, 1735 NEW YORK AVENUE, N.W., WASHINGTON, D.C. 20006

work, and shall properly connect and coordinate his Work with theirs.

**6.2.2** If any part of the Contractor's Work depends for proper execution or results upon the work of any other separate contractor, the Contractor shall inspect and promptly report to the Architect any apparent discrepancies or defects in such work that render it unsuitable for such proper execution and results. Failure of the Contractor so to inspect and report shall constitute an acceptance of the other contractor's work as fit and proper to receive his Work, except as to defects which may develop in the other separate contractor's work after the execution of the Contractor's Work.

**6.2.3** Should the Contractor cause damage to the work or property of any separate contractor on the Project, the Contractor shall, upon due notice, settle with such other contractor by agreement or arbitration, if he will so settle. If such separate contractor sues the Owner or initiates an arbitration proceeding on account of any damage alleged to have been so sustained, the Owner shall notify the Contractor who shall defend such proceedings at the Owner's expense, and if any judgment or award against the Owner arises therefrom the Contractor shall pay or satisfy it and shall reimburse the Owner for all attorneys' fees and court or arbitration costs which the Owner has incurred.

**6.3 CUTTING AND PATCHING**
**UNDER SEPARATE CONTRACTS**

**6.3.1** The Contractor shall be responsible for any cutting, fitting and patching that may be required to complete his Work except as otherwise specifically provided in the Contract Documents. The Contractor shall not endanger any work of any other contractors by cutting, excavating or otherwise altering any work and shall not cut or alter the work of any other contractor except with the written consent of the Architect.

**6.3.2** Any costs caused by defective or ill-timed work shall be borne by the party responsible therefor.

**6.4 OWNER'S RIGHT TO CLEAN UP**

**6.4.1** If a dispute arises between the separate contractors as to their responsibility for cleaning up as required by Paragraph 4.16, the Owner may clean up and charge the cost thereof to the several contractors as the Architect shall determine to be just.

## ARTICLE 7

### MISCELLANEOUS PROVISIONS

**7.1 GOVERNING LAW**

**7.1.1** The Contract shall be governed by the law of the place where the Project is located.

**7.2 SUCCESSORS AND ASSIGNS**

**7.2.1** The Owner and the Contractor each binds himself, his partners, successors, assigns and legal representatives to the other party hereto and to the partners, successors, assigns and legal representatives of such other party in respect to all covenants, agreements and obligations contained in the Contract Documents. Neither

party to the Contract shall assign the Contract or sublet it as a whole without the written consent of the other, nor shall the Contractor assign any moneys due or to become due to him hereunder, without the previous written consent of the Owner.

**7.3 WRITTEN NOTICE**

**7.3.1** Written notice shall be deemed to have been duly served if delivered in person to the individual or member of the firm or to an officer of the corporation for whom it was intended, or if delivered at or sent by registered or certified mail to the last business address known to him who gives the notice.

**7.4 CLAIMS FOR DAMAGES**

**7.4.1** Should either party to the Contract suffer injury or damage to person or property because of any act or omission of the other party or of any of his employees, agents or others for whose acts he is legally liable, claim shall be made in writing to such other party within a reasonable time after the first observance of such injury or damage.

**7.5 PERFORMANCE BOND AND**
**LABOR AND MATERIAL PAYMENT BOND**

**7.5.1** The Owner shall have the right to require the Contractor to furnish bonds covering the faithful performance of the Contract and the payment of all obligations arising thereunder if and as required in the Instructions to Bidders or elsewhere in the Contract Documents.

**7.6 RIGHTS AND REMEDIES**

**7.6.1** The duties and obligations imposed by the Contract Documents and the rights and remedies available thereunder shall be in addition to and not a limitation of any duties, obligations, rights and remedies otherwise imposed or available by law.

**7.7 ROYALTIES AND PATENTS**

**7.7.1** The Contractor shall pay all royalties and license fees. He shall defend all suits or claims for infringement of any patent rights and shall save the Owner harmless from loss on account thereof, except that the Owner shall be responsible for all such loss when a particular design, process or the product of a particular manufacturer or manufacturers is specified, but if the Contractor has reason to believe that the design, process or product specified is an infringement of a patent, he shall be responsible for such loss unless he promptly gives such information to the Architect.

**7.8 TESTS**

**7.8.1** If the Contract Documents, laws, ordinances, rules, regulations or orders of any public authority having jurisdiction require any Work to be inspected, tested or approved, the Contractor shall give the Architect timely notice of its readiness and of the date arranged so the Architect may observe such inspection, testing or approval. The Contractor shall bear all costs of such inspections, tests and approvals unless otherwise provided.

**7.8.2** If after the commencement of the Work the Architect determines that any Work requires special inspection, testing, or approval which Subparagraph 7.8.1

AIA DOCUMENT A201 • GENERAL CONDITIONS OF THE CONTRACT FOR CONSTRUCTION • TWELFTH EDITION • APRIL 1970 ED.
AIA® • © 1970 • THE AMERICAN INSTITUTE OF ARCHITECTS, 1735 NEW YORK AVENUE, N.W., WASHINGTON, D.C. 20006

**11**

does not include, he will, upon written authorization from the Owner, instruct the Contractor to order such special inspection, testing or approval, and the Contractor shall give notice as in Subparagraph 7.8.1. If such special inspection or testing reveals a failure of the Work to comply (1) with the requirements of the Contract Documents or (2), with respect to the performance of the Work, with laws, ordinances, rules, regulations or orders of any public authority having jurisdiction, the Contractor shall bear all costs thereof, including the Architect's additional services made necessary by such failure; otherwise the Owner shall bear such costs, and an appropriate Change Order shall be issued.

**7.8.3** Required certificates of inspection, testing or approval shall be secured by the Contractor and promptly delivered by him to the Architect.

**7.8.4** If the Architect wishes to observe the inspections, tests or approvals required by this Paragraph 7.8, he will do so promptly and, where practicable, at the source of supply.

**7.8.5** Neither the observations of the Architect in his Administration of the Construction Contract, nor inspections, tests or approvals by persons other than the Contractor shall relieve the Contractor from his obligations to perform the Work in accordance with the Contract Documents.

**7.9   INTEREST**
**7.9.1** Any moneys not paid when due to either party under this Contract shall bear interest at the legal rate in force at the place of the Project.

**7.10   ARBITRATION**
**7.10.1** All claims, disputes and other matters in question arising out of, or relating to, this Contract or the breach thereof, except as set forth in Subparagraph 2.2.9 with respect to the Architect's decisions on matters relating to artistic effect, and except for claims which have been waived by the making or acceptance of final payment as provided by Subparagraphs 9.7.5 and 9.7.6, shall be decided by arbitration in accordance with the Construction Industry Arbitration Rules of the American Arbitration Association then obtaining unless the parties mutually agree otherwise. This agreement to arbitrate shall be specifically enforceable under the prevailing arbitration law. The award rendered by the arbitrators shall be final, and judgment may be entered upon it in accordance with applicable law in any court having jurisdiction thereof.

**7.10.2** Notice of the demand for arbitration shall be filed in writing with the other party to the Contract and with the American Arbitration Association, and a copy shall be filed with the Architect. The demand for arbitration shall be made within the time limits specified in Subparagraphs 2.2.10 and 2.2.11 where applicable, and in all other cases within a reasonable time after the claim, dispute or other matter in question has arisen, and in no event shall it be made after the date when institution of legal or equitable proceedings based on such claim, dispute or other matter in question would be barred by the applicable statute of limitations.

**7.10.3** The Contractor shall carry on the Work and maintain the progress schedule during any arbitration proceedings, unless otherwise agreed by him and the Owner in writing.

# ARTICLE 8

## TIME

**8.1   DEFINITIONS**
**8.1.1** The Contract Time is the period of time allotted in the Contract Documents for completion of the Work.

**8.1.2** The date of commencement of the Work is the date established in a notice to proceed. If there is no notice to proceed, it shall be the date of the Agreement or such other date as may be established therein.

**8.1.3** The Date of Substantial Completion of the Work or designated portion thereof is the Date certified by the Architect when construction is sufficiently complete, in accordance with the Contract Documents, so the Owner may occupy the Work or designated portion thereof for the use for which it is intended.

**8.1.4** The term day as used in the Contract Documents shall mean calendar day.

**8.2   PROGRESS AND COMPLETION**
**8.2.1** All time limits stated in the Contract Documents are of the essence of the Contract

**8.2.2** The Contractor shall begin the Work on the date of commencement as defined in Subparagraph 8.1.2. He shall carry the Work forward expeditiously with adequate forces and shall complete it within the Contract Time.

**8.2.3** If a date or time of completion is included in the Contract, it shall be the Date of Substantial Completion as defined in Subparagraph 8.1.3, including authorized extensions thereto, unless otherwise provided.

**8.3   DELAYS AND EXTENSIONS OF TIME**
**8.3.1** If the Contractor is delayed at any time in the progress of the Work by any act or neglect of the Owner or the Architect, or by any employee of either, or by any separate contractor employed by the Owner, or by changes ordered in the Work, or by labor disputes, fire, unusual delay in transportation, unavoidable casualties or any causes beyond the Contractor's control, or by delay authorized by the Owner pending arbitration, or by any cause which the Architect determines may justify the delay, then the Contract Time shall be extended by Change Order for such reasonable time as the Architect may determine.

**8.3.2** All claims for extension of time shall be made in writing to the Architect no more than twenty days after the occurrence of the delay; otherwise they shall be waived. In the case of a continuing cause of delay only one claim is necessary.

**8.3.3** If no schedule or agreement is made stating the dates upon which written interpretations as set forth in Subparagraph 1.2.5 shall be furnished, then no claim for delay shall be allowed on account of failure to furnish

**AIA DOCUMENT A201** • GENERAL CONDITIONS OF THE CONTRACT FOR CONSTRUCTION • TWELFTH EDITION • APRIL 1970 ED.
AIA® • © 1970 • THE AMERICAN INSTITUTE OF ARCHITECTS, 1735 NEW YORK AVENUE, N.W., WASHINGTON, D.C. 20006

such interpretations until fifteen days after demand is made for them, and not then unless such claim is reasonable.

**8.3.4** This Paragraph 8.3 does not exclude the recovery of damages for delay by either party under other provisions of the Contract Documents.

# ARTICLE 9

## PAYMENTS AND COMPLETION

**9.1    CONTRACT SUM**

**9.1.1** The Contract Sum is stated in the Agreement and is the total amount payable by the Owner to the Contractor for the performance of the Work under the Contract Documents.

**9.2    SCHEDULE OF VALUES**

**9.2.1** Before the first Application for Payment, the Contractor shall submit to the Architect a schedule of values of the various portions of the Work, including quantities if required by the Architect, aggregating the total Contract Sum, divided so as to facilitate payments to Subcontractors in accordance with Paragraph 5.4, prepared in such form as specified or as the Architect and the Contractor may agree upon, and supported by such data to substantiate its correctness as the Architect may require. Each item in the schedule of values shall include its proper share of overhead and profit. This schedule, when approved by the Architect, shall be used only as a basis for the Contractor's Applications for Payment.

**9.3    PROGRESS PAYMENTS**

**9.3.1** At least ten days before each progress payment falls due, the Contractor shall submit to the Architect an itemized Application for Payment, supported by such data substantiating the Contractor's right to payment as the Owner or the Architect may require.

**9.3.2** If payments are to be made on account of materials or equipment not incorporated in the Work but delivered and suitably stored at the site, or at some other location agreed upon in writing, such payments shall be conditioned upon submission by the Contractor of bills of sale or such other procedures satisfactory to the Owner to establish the Owner's title to such materials or equipment or otherwise protect the Owner's interest including applicable insurance and transportation to the site.

**9.3.3** The Contractor warrants and guarantees that title to all Work, materials and equipment covered by an Application for Payment, whether incorporated in the Project or not, will pass to the Owner upon the receipt of such payment by the Contractor, free and clear of all liens, claims, security interests or encumbrances, hereinafter referred to in this Article 9 as "liens"; and that no Work, materials or equipment covered by an Application for Payment will have been acquired by the Contractor or by any other person performing the Work at the site or furnishing materials and equipment for the Project, subject to an agreement under which an interest therein or an encumbrance thereon is retained by the seller or otherwise imposed by the Contractor or such other person.

**9.4    CERTIFICATES FOR PAYMENT**

**9.4.1** If the Contractor has made Application for Payment as above, the Architect will, with reasonable promptness but not more than seven days after the receipt of the Application, issue a Certificate for Payment to the Owner, with a copy to the Contractor, for such amount as he determines to be properly due, or state in writing his reasons for withholding a Certificate as provided in Subparagraph 9.5.1.

**9.4.2** The issuance of a Certificate for Payment will constitute a representation by the Architect to the Owner, based on his observations at the site as provided in Subparagraph 2.2.4 and the data comprising the Application for Payment, that the Work has progressed to the point indicated; that, to the best of his knowledge, information and belief, the quality of the Work is in accordance with the Contract Documents (subject to an evaluation of the Work for conformance with the Contract Documents upon Substantial Completion, to the results of any subsequent tests required by the Contract Documents, to minor deviations from the Contract Documents correctable prior to completion, and to any specific qualifications stated in his Certificate); and that the Contractor is entitled to payment in the amount certified. In addition, the Architect's final Certificate for Payment will constitute a further representation that the conditions precedent to the Contractor's being entitled to final payment as set forth in Subparagraph 9.7.2 have been fulfilled. However, by issuing a Certificate for Payment, the Architect shall not thereby be deemed to represent that he has made exhaustive or continuous on-site inspections to check the quality or quantity of the Work or that he has reviewed the construction means, methods, techniques, sequences or procedures, or that he has made any examination to ascertain how or for what purpose the Contractor has used the moneys previously paid on account of the Contract Sum.

**9.4.3** After the Architect has issued a Certificate for Payment, the Owner shall make payment in the manner provided in the Agreement.

**9.4.4** No certificate for a progress payment, nor any progress payment, nor any partial or entire use or occupancy of the Project by the Owner, shall constitute an acceptance of any Work not in accordance with the Contract Documents.

**9.5    PAYMENTS WITHHELD**

**9.5.1** The Architect may decline to approve an Application for Payment and may withhold his Certificate in whole or in part, to the extent necessary reasonably to protect the Owner, if in his opinion he is unable to make representations to the Owner as provided in Subparagraph 9.4.2. The Architect may also decline to approve any Applications for Payment or, because of subsequently discovered evidence or subsequent inspections, he may nullify the whole or any part of any Certificate for Payment previously issued, to such extent as may be necessary in his opinion to protect the Owner from loss because of:

.1 defective work not remedied,

.2 third party claims filed or reasonable evidence indicating probable filing of such claims,

AIA DOCUMENT A201 • GENERAL CONDITIONS OF THE CONTRACT FOR CONSTRUCTION • TWELFTH EDITION • APRIL 1970 ED.
AIA® • © 1970 • THE AMERICAN INSTITUTE OF ARCHITECTS, 1735 NEW YORK AVENUE, N.W., WASHINGTON, D.C. 20006    **13**

.3 failure of the Contractor to make payments properly to Subcontractors or for labor, materials or equipment,

.4 reasonable doubt that the Work can be completed for the unpaid balance of the Contract Sum,

.5 damage to another contractor,

.6 reasonable indication that the Work will not be completed within the Contract Time, or

.7 unsatisfactory prosecution of the Work by the Contractor.

**9.5.2** When the above grounds in Subparagraph 9.5.1 are removed, payment shall be made for amounts withheld because of them.

**9.6  FAILURE OF PAYMENT**

**9.6.1** If the Architect should fail to issue any Certificate for Payment, through no fault of the Contractor, within seven days after receipt of the Contractor's Application for Payment, or if the Owner should fail to pay the Contractor within seven days after the date of payment established in the Agreement any amount certified by the Architect or awarded by arbitration, then the Contractor may, upon seven additional days' written notice to the Owner and the Architect, stop the Work until payment of the amount owing has been received.

**9.7  SUBSTANTIAL COMPLETION AND FINAL PAYMENT**

**9.7.1** When the Contractor determines that the Work or a designated portion thereof acceptable to the Owner is substantially complete, the Contractor shall prepare for submission to the Architect a list of items to be completed or corrected. The failure to include any items on such list does not alter the responsibility of the Contractor to complete all Work in accordance with the Contract Documents. When the Architect on the basis of an inspection determines that the Work is substantially complete, he will then prepare a Certificate of Substantial Completion which shall establish the Date of Substantial Completion, shall state the responsibilities of the Owner and the Contractor for maintenance, heat, utilities, and insurance, and shall fix the time within which the Contractor shall complete the items listed therein. The Certificate of Substantial Completion shall be submitted to the Owner and the Contractor for their written acceptance of the responsibilities assigned to them in such Certificate.

**9.7.2** Upon receipt of written notice that the Work is ready for final inspection and acceptance and upon receipt of a final Application for Payment, the Architect will promptly make such inspection and, when he finds the Work acceptable under the Contract Documents and the Contract fully performed, he will promptly issue a final Certificate for Payment stating that to the best of his knowledge, information and belief, and on the basis of his observations and inspections, the Work has been completed in accordance with the terms and conditions of the Contract Documents and that the entire balance found to be due the Contractor, and noted in said final Certificate, is due and payable.

**9.7.3** Neither the final payment nor the remaining retained percentage shall become due until the Contractor

submits to the Architect (1) an Affidavit that all payrolls, bills for materials and equipment, and other indebtedness connected with the Work for which the Owner or his property might in any way be responsible, have been paid or otherwise satisfied, (2) consent of surety, if any, to final payment and (3), if required by the Owner, other data establishing payment or satisfaction of all such obligations, such as receipts, releases and waivers of liens arising out of the Contract, to the extent and in such form as may be designated by the Owner. If any Subcontractor refuses to furnish a release or waiver required by the Owner, the Contractor may furnish a bond satisfactory to the Owner to indemnify him against any such lien. If any such lien remains unsatisfied after all payments are made, the Contractor shall refund to the Owner all moneys that the latter may be compelled to pay in discharging such lien, including all costs and reasonable attorneys' fees.

**9.7.4** If after Substantial Completion of the Work final completion thereof is materially delayed through no fault of the Contractor, and the Architect so confirms, the Owner shall, upon certification by the Architect, and without terminating the Contract, make payment of the balance due for that portion of the Work fully completed and accepted. If the remaining balance for Work not fully completed or corrected is less than the retainage stipulated in the Agreement, and if bonds have been furnished as required in Subparagraph 7.5.1, the written consent of the surety to the payment of the balance due for that portion of the Work fully completed and accepted shall be submitted by the Contractor to the Architect prior to certification of such payment. Such payment shall be made under the terms and conditions governing final payment, except that it shall not constitute a waiver of claims.

**9.7.5** The making of final payment shall constitute a waiver of all claims by the Owner except those arising from:

.1 unsettled liens,

.2 faulty or defective Work appearing after Substantial Completion,

.3 failure of the Work to comply with the requirements of the Contract Documents, or

.4 terms of any special guarantees required by the Contract Documents.

**9.7.6** The acceptance of final payment shall constitute a waiver of all claims by the Contractor except those previously made in writing and still unsettled.

## ARTICLE 10

## PROTECTION OF PERSONS AND PROPERTY

**10.1  SAFETY PRECAUTIONS AND PROGRAMS**

**10.1.1** The Contractor shall be responsible for initiating, maintaining and supervising all safety precautions and programs in connection with the Work.

**10.2  SAFETY OF PERSONS AND PROPERTY**

**10.2.1** The Contractor shall take all reasonable precautions for the safety of, and shall provide all reasonable protection to prevent damage, injury or loss to:

AIA DOCUMENT A201 • GENERAL CONDITIONS OF THE CONTRACT FOR CONSTRUCTION • TWELFTH EDITION • APRIL 1970 ED. AIA® • © 1970 • THE AMERICAN INSTITUTE OF ARCHITECTS, 1735 NEW YORK AVENUE, N.W., WASHINGTON, D.C. 20006

.1 all employees on the Work and all other persons who may be affected thereby;

.2 all the Work and all materials and equipment to be incorporated therein, whether in storage on or off the site, under the care, custody or control of the Contractor or any of his Subcontractors or Sub-subcontractors; and

.3 other property at the site or adjacent thereto, including trees, shrubs, lawns, walks, pavements, roadways, structures and utilities not designated for removal, relocation or replacement in the course of construction.

**10.2.2** The Contractor shall comply with all applicable laws, ordinances, rules, regulations and lawful orders of any public authority having jurisdiction for the safety of persons or property or to protect them from damage, injury or loss. He shall erect and maintain, as required by existing conditions and progress of the Work, all reasonable safeguards for safety and protection, including posting danger signs and other warnings against hazards, promulgating safety regulations and notifying owners and users of adjacent utilities.

**10.2.3** When the use or storage of explosives or other hazardous materials or equipment is necessary for the execution of the Work, the Contractor shall exercise the utmost care and shall carry on such activities under the supervision of properly qualified personnel.

**10.2.4** All damage or loss to any property referred to in Clauses 10.2.1.2 and 10.2.1.3 caused in whole or in part by the Contractor, any Subcontractor, any Sub-subcontractor, or anyone directly or indirectly employed by any of them, or by anyone for whose acts any of them may be liable, shall be remedied by the Contractor, except damage or loss attributable to faulty Drawings or Specifications or to the acts or omissions of the Owner or Architect or anyone employed by either of them or for whose acts either of them may be liable, and not attributable to the fault or negligence of the Contractor.

**10.2.5** The Contractor shall designate a responsible member of his organization at the site whose duty shall be the prevention of accidents. This person shall be the Contractor's superintendent unless otherwise designated in writing by the Contractor to the Owner and the Architect.

**10.2.6** The Contractor shall not load or permit any part of the Work to be loaded so as to endanger its safety.

**10.3  EMERGENCIES**

**10.3.1** In any emergency affecting the safety of persons or property, the Contractor shall act, at his discretion, to prevent threatened damage, injury or loss. Any additional compensation or extension of time claimed by the Contractor on account of emergency work shall be determined as provided in Article 12 for Changes in the Work.

## ARTICLE 11

### INSURANCE

**11.1  CONTRACTOR'S LIABILITY INSURANCE**

**11.1.1** The Contractor shall purchase and maintain such insurance as will protect him from claims set forth below which may arise out of or result from the Contractor's operations under the Contract, whether such operations be by himself or by any Subcontractor or by anyone directly or indirectly employed by any of them, or by anyone for whose acts any of them may be liable:

.1 claims under workmen's compensation, disability benefit and other similar employee benefit acts;

.2 claims for damages because of bodily injury, occupational sickness or disease, or death of his employees;

.3 claims for damages because of bodily injury, sickness or disease, or death of any person other than his employees;

.4 claims for damages insured by usual personal injury liability coverage which are sustained (1) by any person as a result of an offense directly or indirectly related to the employment of such person by the Contractor, or (2) by any other person; and

.5 claims for damages because of injury to or destruction of tangible property, including loss of use resulting therefrom.

**11.1.2** The insurance required by Subparagraph 11.1.1 shall be written for not less than any limits of liability specified in the Contract Documents, or required by law, whichever is greater, and shall include contractual liability insurance as applicable to the Contractor's obligations under Paragraph 4.18.

**11.1.3** Certificates of Insurance acceptable to the Owner shall be filed with the Owner prior to commencement of the Work. These Certificates shall contain a provision that coverages afforded under the policies will not be cancelled until at least fifteen days' prior written notice has been given to the Owner.

**11.2  OWNER'S LIABILITY INSURANCE**

**11.2.1** The Owner shall be responsible for purchasing and maintaining his own liability insurance and, at his option, may purchase and maintain such insurance as will protect him against claims which may arise from operations under the Contract.

**11.3  PROPERTY INSURANCE**

**11.3.1** Unless otherwise provided, the Owner shall purchase and maintain property insurance upon the entire Work at the site to the full insurable value thereof. This insurance shall include the interests of the Owner, the Contractor, Subcontractors and Sub-subcontractors in the Work and shall insure against the perils of Fire, Extended Coverage, Vandalism and Malicious Mischief.

**11.3.2** The Owner shall purchase and maintain such steam boiler and machinery insurance as may be required by the Contract Documents or by law. This insurance shall include the interests of the Owner, the Contractor, Subcontractors and Sub-subcontractors in the Work.

**11.3.3** Any insured loss is to be adjusted with the Owner and made payable to the Owner as trustee for the insureds, as their interests may appear, subject to the requirements of any applicable mortgagee clause and of Subparagraph 11.3.8.

**11.3.4** The Owner shall file a copy of all policies with the Contractor before an exposure to loss may occur. If the Owner does not intend to purchase such insurance, he shall inform the Contractor in writing prior to commencement of the Work. The Contractor may then effect insurance which will protect the interests of himself, his Subcontractors and the Sub-subcontractors in the Work, and by appropriate Change Order the cost thereof shall be charged to the Owner. If the Contractor is damaged by failure of the Owner to purchase or maintain such insurance and so to notify the Contractor, then the Owner shall bear all reasonable costs properly attributable thereto.

**11.3.5** If the Contractor requests in writing that insurance for special hazards be included in the property insurance policy, the Owner shall, if possible, include such insurance, and the cost thereof shall be charged to the Contractor by appropriate Change Order.

**11.3.6** The Owner and Contractor waive all rights against each other for damages caused by fire or other perils to the extent covered by insurance provided under this Paragraph 11.3, except such rights as they may have to the proceeds of such insurance held by the Owner as trustee. The Contractor shall require similar waivers by Subcontractors and Sub-subcontractors in accordance with Clause 5.3.1.5.

**11.3.7** If required in writing by any party in interest, the Owner as trustee shall, upon the occurrence of an insured loss, give bond for the proper performance of his duties. He shall deposit in a separate account any money so received, and he shall distribute it in accordance with such agreement as the parties in interest may reach, or in accordance with an award by arbitration in which case the procedure shall be as provided in Paragraph 7.10. If after such loss no other special agreement is made, replacement of damaged work shall be covered by an appropriate Change Order.

**11.3.8** The Owner as trustee shall have power to adjust and settle any loss with the insurers unless one of the parties in interest shall object in writing within five days after the occurrence of loss to the Owner's exercise of this power, and if such objection be made, arbitrators shall be chosen as provided in Paragraph 7.10. The Owner as trustee shall, in that case, make settlement with the insurers in accordance with the directions of such arbitrators. If distribution of the insurance proceeds by arbitration is required, the arbitrators will direct such distribution.

**11.4  LOSS OF USE INSURANCE**

**11.4.1** The Owner, at his option, may purchase and maintain such insurance as will insure him against loss of use of his property due to fire or other hazards, however caused.

## ARTICLE 12

## CHANGES IN THE WORK

**12.1  CHANGE ORDERS**

**12.1.1** The Owner, without invalidating the Contract, may order Changes in the Work within the general scope of the Contract consisting of additions, deletions or other revisions, the Contract Sum and the Contract Time being adjusted accordingly. All such Changes in the Work shall be authorized by Change Order, and shall be executed under the applicable conditions of the Contract Documents.

**12.1.2** A Change Order is a written order to the Contractor signed by the Owner and the Architect, issued after the execution of the Contract, authorizing a Change in the Work or an adjustment in the Contract Sum or the Contract Time. Alternatively, the Change Order may be signed by the Architect alone, provided he has written authority from the Owner for such procedure and that a copy of such written authority is furnished to the Contractor upon request. A Change Order may also be signed by the Contractor if he agrees to the adjustment in the Contract Sum or the Contract Time. The Contract Sum and the Contract Time may be changed only by Change Order.

**12.1.3** The cost or credit to the Owner resulting from a Change in the Work shall be determined in one or more of the following ways:

.1 by mutual acceptance of a lump sum properly itemized;

.2 by unit prices stated in the Contract Documents or subsequently agreed upon; or

.3 by cost and a mutually acceptable fixed or percentage fee.

**12.1.4** If none of the methods set forth in Subparagraph 12.1.3 is agreed upon, the Contractor, provided he receives a Change Order, shall promptly proceed with the Work involved. The cost of such Work shall then be determined by the Architect on the basis of the Contractor's reasonable expenditures and savings, including, in the case of an increase in the Contract Sum, a reasonable allowance for overhead and profit. In such case, and also under Clause 12.1.3.3 above, the Contractor shall keep and present, in such form as the Architect may prescribe, an itemized accounting together with appropriate supporting data. Pending final determination of cost to the Owner, payments on account shall be made on the Architect's Certificate for Payment. The amount of credit to be allowed by the Contractor to the Owner for any deletion or change which results in a net decrease in cost will be the amount of the actual net decrease as confirmed by the Architect. When both additions and credits are involved in any one change, the allowance for overhead and profit shall be figured on the basis of net increase, if any.

**12.1.5** If unit prices are stated in the Contract Documents or subsequently agreed upon, and if the quantities originally contemplated are so changed in a proposed Change Order that application of the agreed unit prices to the quantities of Work proposed will create a hardship on the Owner or the Contractor, the applicable unit prices shall be equitably adjusted to prevent such hardship.

**12.1.6** Should concealed conditions encountered in the performance of the Work below the surface of the ground be at variance with the conditions indicated by the Contract Documents or should unknown physical conditions below the surface of the ground of an unusual nature,

AIA DOCUMENT A201 • GENERAL CONDITIONS OF THE CONTRACT FOR CONSTRUCTION • TWELFTH EDITION • APRIL 1970 ED.
AIA® • © 1970 • THE AMERICAN INSTITUTE OF ARCHITECTS, 1735 NEW YORK AVENUE, N.W., WASHINGTON, D.C. 20006

differing materially from those ordinarily encountered and generally recognized as inherent in work of the character provided for in this Contract, be encountered, the Contract Sum shall be equitably adjusted by Change Order upon claim by either party made within twenty days after the first observance of the conditions.

**12.1.7** If the Contractor claims that additional cost is involved because of (1) any written interpretation issued pursuant to Subparagraph 1.2.5, (2) any order by the Owner to stop the Work pursuant to Paragraph 3.3 where the Contractor was not at fault, or (3) any written order for a minor change in the Work issued pursuant to Paragraph 12.3, the Contractor shall make such claim as provided in Paragraph 12.2.

**12.2   CLAIMS FOR ADDITIONAL COST**

**12.2.1** If the Contractor wishes to make a claim for an increase in the Contract Sum, he shall give the Architect written notice thereof within twenty days after the occurrence of the event giving rise to such claim. This notice shall be given by the Contractor before proceeding to execute the Work, except in an emergency endangering life or property in which case the Contractor shall proceed in accordance with Subparagraph 10.3.1. No such claim shall be valid unless so made. If the Owner and the Contractor cannot agree on the amount of the adjustment in the Contract Sum, it shall be determined by the Architect. Any change in the Contract Sum resulting from such claim shall be authorized by Change Order.

**12.3   MINOR CHANGES IN THE WORK**

**12.3.1** The Architect shall have authority to order minor changes in the Work not involving an adjustment in the Contract Sum or an extension of the Contract Time and not inconsistent with the intent of the Contract Documents. Such changes may be effected by Field Order or by other written order. Such changes shall be binding on the Owner and the Contractor.

**12.4   FIELD ORDERS**

**12.4.1** The Architect may issue written Field Orders which interpret the Contract Documents in accordance with Subparagraph 1.2.5 or which order minor changes in the Work in accordance with Paragraph 12.3 without change in Contract Sum or Contract Time. The Contractor shall carry out such Field Orders promptly.

# ARTICLE 13

## UNCOVERING AND CORRECTION OF WORK

**13.1   UNCOVERING OF WORK**

**13.1.1** If any Work should be covered contrary to the request of the Architect, it must, if required by the Architect, be uncovered for his observation and replaced, at the Contractor's expense.

**13.1.2** If any other Work has been covered which the Architect has not specifically requested to observe prior to being covered, the Architect may request to see such Work and it shall be uncovered by the Contractor. If such Work be found in accordance with the Contract Documents, the cost of uncovering and replacement

shall, by appropriate Change Order, be charged to the Owner. If such Work be found not in accordance with the Contract Documents, the Contractor shall pay such costs unless it be found that this condition was caused by a separate contractor employed as provided in Article 6, and in that event the Owner shall be responsible for the payment of such costs.

**13.2   CORRECTION OF WORK**

**13.2.1** The Contractor shall promptly correct all Work rejected by the Architect as defective or as failing to conform to the Contract Documents whether observed before or after Substantial Completion and whether or not fabricated, installed or completed. The Contractor shall bear all cost of correcting such rejected Work, including the cost of the Architect's additional services thereby made necessary.

**13.2.2** If, within one year after the Date of Substantial Completion or within such longer period of time as may be prescribed by law or by the terms of any applicable special guarantee required by the Contract Documents, any of the Work is found to be defective or not in accordance with the Contract Documents, the Contractor shall correct it promptly after receipt of a written notice from the Owner to do so unless the Owner has previously given the Contractor a written acceptance of such condition. The Owner shall give such notice promptly after discovery of the condition.

**13.2.3** All such defective or non-conforming Work under Subparagraphs 13.2.1 and 13.2.2 shall be removed from the site if necessary, and the Work shall be corrected to comply with the Contract Documents without cost to the Owner.

**13.2.4** The Contractor shall bear the cost of making good all work of separate contractors destroyed or damaged by such removal or correction.

**13.2.5** If the Contractor does not remove such defective or non-conforming Work within a reasonable time fixed by written notice from the Architect, the Owner may remove it and may store the materials or equipment at the expense of the Contractor. If the Contractor does not pay the cost of such removal and storage within ten days thereafter, the Owner may upon ten additional days' written notice sell such Work at auction or at private sale and shall account for the net proceeds thereof, after deducting all the costs that should have been borne by the Contractor including compensation for additional architectural services. If such proceeds of sale do not cover all costs which the Contractor should have borne, the difference shall be charged to the Contractor and an appropriate Change Order shall be issued. If the payments then or thereafter due the Contractor are not sufficient to cover such amount, the Contractor shall pay the difference to the Owner.

**13.2.6** If the Contractor fails to correct such defective or non-conforming Work, the Owner may correct it in accordance with Paragraph 3.4.

**13.3   ACCEPTANCE OF DEFECTIVE
OR NON-CONFORMING WORK**

**13.3.1** If the Owner prefers to accept defective or nonconforming Work, he may do so instead of requiring its removal and correction, in which case a Change Order will be issued to reflect an appropriate reduction in the Contract Sum, or, if the amount is determined after final payment, it shall be paid by the Contractor.

# ARTICLE 14

# TERMINATION OF THE CONTRACT

## 14.1 TERMINATION BY THE CONTRACTOR

**14.1.1** If the Work is stopped for a period of thirty days under an order of any court or other public authority having jurisdiction, or as a result of an act of government, such as a declaration of a national emergency making materials unavailable, through no act or fault of the Contractor or a Subcontractor or their agents or employees or any other persons performing any of the Work under a contract with the Contractor, or if the Work should be stopped for a period of thirty days by the Contractor for the Architect's failure to issue a Certificate for Payment as provided in Paragraph 9.6 or for the Owner's failure to make payment thereon as provided in Paragraph 9.6, then the Contractor may, upon seven days' written notice to the Owner and the Architect, terminate the Contract and recover from the Owner payment for all Work executed and for any proven loss sustained upon any materials, equipment, tools, construction equipment and machinery, including reasonable profit and damages.

## 14.2 TERMINATION BY THE OWNER

**14.2.1** If the Contractor is adjudged a bankrupt, or if he makes a general assignment for the benefit of his creditors, or if a receiver is appointed on account of his insolvency, or if he persistently or repeatedly refuses or fails, except in cases for which extension of time is provided, to supply enough properly skilled workmen or proper materials, or if he fails to make prompt payment to Subcontractors or for materials or labor, or persistently disregards laws, ordinances, rules, regulations or orders of any public authority having jurisdiction, or otherwise is guilty of a substantial violation of a provision of the Contract Documents, then the Owner, upon certification by the Architect that sufficient cause exists to justify such action, may, without prejudice to any right or remedy and after giving the Contractor and his surety, if any, seven days' written notice, terminate the employment of the Contractor and take possession of the site and of all materials, equipment, tools, construction equipment and machinery thereon owned by the Contractor and may finish the Work by whatever method he may deem expedient. In such case the Contractor shall not be entitled to receive any further payment until the Work is finished.

**14.2.2** If the unpaid balance of the Contract Sum exceeds the costs of finishing the Work, including compensation for the Architect's additional services, such excess shall be paid to the Contractor. If such costs exceed such unpaid balance, the Contractor shall pay the difference to the Owner. The costs incurred by the Owner as herein provided shall be certified by the Architect.

**AIA DOCUMENT A201** • GENERAL CONDITIONS OF THE CONTRACT FOR CONSTRUCTION • TWELFTH EDITION • APRIL 1970 ED.
AIA® • © 1970 • THE AMERICAN INSTITUTE OF ARCHITECTS, 1735 NEW YORK AVENUE, N.W., WASHINGTON, D.C. 20006

# THE AMERICAN INSTITUTE OF ARCHITECTS

AIA Document A107

# Standard Form of Agreement Between Owner and Contractor

## Short Form Agreement for **Small Construction Contracts**

*Where the Basis of Payment is a*

### STIPULATED SUM

THIS DOCUMENT HAS IMPORTANT LEGAL CONSEQUENCES; CONSULTATION WITH
AN ATTORNEY IS ENCOURAGED WITH RESPECT TO ITS COMPLETION OR MODIFICATION

For other contracts the AIA issues Standard Forms of Owner-Contractor Agreements and Standard General Conditions
of the Contract for Construction for use in connection therewith.

---

**AGREEMENT**

made this                    day of                    in the year Nineteen
Hundred and

**BETWEEN**

the Owner, and

the Contractor.

"This document has been reproduced with the permission of
The American Institute of Architects. Further reproduction is
not authorized".

The Owner and Contractor agree as set forth below.

---

**AIA DOCUMENT A107** • SMALL CONSTRUCTION CONTRACT • SEPTEMBER 1970 EDITION • AIA®
©1970 • THE AMERICAN INSTITUTE OF ARCHITECTS, 1735 NEW YORK AVE., N.W., WASHINGTON, D.C. 20006

1

# ARTICLE 1
## THE WORK

The Contractor shall perform all the Work required by the Contract Documents for
*(Here insert the caption descriptive of the Work as used on other Contract Documents.)*


# ARTICLE 2
## ARCHITECT

The Architect for this Project is


# ARTICLE 3
## TIME OF COMMENCEMENT AND COMPLETION

The Work to be performed under this Contract shall be commenced


and completed


# ARTICLE 4
## CONTRACT SUM

The Owner shall pay the Contractor for the performance of the Work, subject to additions and deductions by Change Order as provided in the General Conditions, in current funds, the Contract Sum of
*(State here the lump sum amount, unit prices, or both, as desired.)*

---

## ARTICLE 5
## PROGRESS PAYMENTS

Based upon Applications for Payment submitted to the Architect by the Contractor and Certificates for Payment issued by the Architect, the Owner shall make progress payments on account of the Contract Sum to the Contractor as follows:

## ARTICLE 6
## FINAL PAYMENT

The Owner shall make final payment                                        days after completion of the Work, provided the Contract be then fully performed, subject to the provisions of Article 17 of the General Conditions.

## ARTICLE 7
## ENUMERATION OF CONTRACT DOCUMENTS

The Contract Documents are as noted in Paragraph 8.1 of the General Conditions and are enumerated as follows:
*(List below the Agreement, Conditions of the Contract (General, Supplementary, and other Conditions), Drawings, Specifications, Addenda and accepted Alternates, showing page or sheet numbers in all cases and dates where applicable.)*

## ARTICLE 8
## CONTRACT DOCUMENTS

**8.1** The Contract Documents consist of this Agreement (which includes the General Conditions), Supplementary and other Conditions, the Drawings, the Specifications, all Addenda issued prior to the execution of this Agreement, all amendments, Change Orders, and written interpretations of the Contract Documents issued by the Architect. These form the Contract and what is required by any one shall be as binding as if required by all. The intention of the Contract Documents is to include all labor, materials, equipment and other items as provided in Paragraph 11.2 necessary for the proper execution and completion of the Work and the terms and conditions of payment therefor, and also to include all Work which may be reasonably inferable from the Contract Documents as being necessary to produce the intended results.

**8.2** The Contract Documents shall be signed in not less than triplicate by the Owner and the Contractor. If either the Owner or the Contractor do not sign the Drawings, Specifications, or any of the other Contract Documents, the Architect shall identify them. By executing the Contract, the Contractor represents that he has visited the site and familiarized himself with the local conditions under which the Work is to be performed.

**.3** The term Work as used in the Contract Documents includes all labor necessary to produce the construction required by the Contract Documents, and all materials and equipment incorporated or to be incorporated in such construction.

## ARTICLE 9
## ARCHITECT

**9.1** The Architect will provide general administration of the Contract and will be the Owner's representative during the construction period.

**9.2** The Architect shall at all times have access to the Work wherever it is in preparation and progress.

**9.3** The Architect will make periodic visits to the site to familiarize himself generally with the progress and quality of the Work and to determine in general if the Work is proceeding in accordance with the Contract Documents. On the basis of his on-site observations as an architect, he will keep the Owner informed of the progress of the Work, and will endeavor to guard the Owner against defects and deficiencies in the Work of the Contractor. The Architect will not be required to make exhaustive or continuous on-site inspections to check the quality or quantity of the Work. The Architect will not be responsible for construction means, methods, techniques, sequences or procedures, or for safety precautions and programs in connection with the Work, and he will not be responsible for the Contractor's failure to carry out the Work in accordance with the Contract Documents.

**9.4** Based on such observations and the Contractor's Applications for Payment, the Architect will determine the amounts owing to the Contractor and will issue Certificates for Payment in accordance with Article 17.

**9.5** The Architect will be, in the first instance, the interpreter of the requirements of the Contract Documents. He will make decisions on all claims and disputes between the Owner and the Contractor. All his decisions are subject to arbitration.

**9.6** The Architect will have authority to reject Work which does not conform to the Contract Documents.

## ARTICLE 10
## OWNER

**10.1** The Owner shall furnish all surveys.

**10.2** The Owner shall secure and pay for easements for permanent structures or permanent changes in existing facilities.

**10.3** The Owner shall issue all instructions to the Contractor through the Architect.

## ARTICLE 11
## CONTRACTOR

**11.1** The Contractor shall supervise and direct the Work, using his best skill and attention. The Contractor shall be solely responsible for all construction means, methods, techniques, sequences and procedures and for coordinating all portions of the Work under the Contract.

**11.2** Unless otherwise specifically noted, the Contractor shall provide and pay for all labor, materials, equipment, tools, construction equipment and machinery, water, heat, utilities, transportation, and other facilities and services necessary for the proper execution and completion of the Work.

**11.3** The Contractor shall at all times enforce strict discipline and good order among his employees, and shall not employ on the Work any unfit person or anyone not skilled in the task assigned to him.

**11.4** The Contractor warrants to the Owner and the Architect that all materials and equipment incorporated in the Work will be new unless otherwise specified, and that all Work will be of good quality, free from faults and defects and in conformance with the Contract Documents. All Work not so conforming to these standards may be considered defective.

**11.5** The Contractor shall pay all sales, consumer, use and other similar taxes required by law and shall secure all permits, fees and licenses necessary for the execution of the Work.

**11.6** The Contractor shall give all notices and comply with all laws, ordinances, rules, regulations, and orders of any public authority bearing on the performance of

AIA DOCUMENT A107 • SMALL CONSTRUCTION CONTRACT • SEPTEMBER 1970 EDITION • AIA®
©1970 • THE AMERICAN INSTITUTE OF ARCHITECTS, 1735 NEW YORK AVE., N.W., WASHINGTON, D.C. 20006

**4**

the Work, and shall notify the Architect if the Drawings and Specifications are at variance therewith.

**11.7** The Contractor shall be responsible for the acts and omissions of all his employees and all Subcontractors, their agents and employees and all other persons performing any of the Work under a contract with the Contractor.

**11.8** The Contractor shall review, stamp with his approval and submit all samples and shop drawings as directed for approval of the Architect for conformance with the design concept and with the information given in the Contract Documents. The Work shall be in accordance with approved samples and shop drawings.

**11.9** The Contractor at all times shall keep the premises free from accumulation of waste materials or rubbish caused by his operations. At the completion of the Work he shall remove all his waste materials and rubbish from and about the Project as well as his tools, construction equipment, machinery and surplus materials, and shall clean all glass surfaces and shall leave the Work "broom clean" or its equivalent, except as otherwise specified.

**11.10** The Contractor shall indemnify and hold harmless the Owner and the Architect and their agents and employees from and against all claims, damages, losses and expenses including attorneys' fees arising out of or resulting from the performance of the Work, provided that any such claim, damage, loss or expense (1) is attributable to bodily injury, sickness, disease or death, or to injury to or destruction of tangible property (other than the Work itself) including the loss of use resulting therefrom, and (2) is caused in whole or in part by any negligent act or omission of the Contractor, any Subcontractor, anyone directly or indirectly employed by any of them or anyone for whose acts any of them may be liable, regardless of whether or not it is caused in part by a party indemnified hereunder. In any and all claims against the Owner or the Architect or any of their agents or employees by any employee of the Contractor, any Subcontractor, anyone directly or indirectly employed by any of them or anyone for whose acts any of them may be liable, the indemnification obligation under this Paragraph 11.10 shall not be limited in any way by any limitation on the amount or type of damages, compensation or benefits payable by or for the Contractor or any Subcontractor under workmen's compensation acts, disability benefit acts or other employee benefit acts. The obligations of the Contractor under this Paragraph 11.10 shall not extend to the liability of the Architect, his agents or employees arising out of (1) the preparation or approval of maps, drawings, opinions, reports, surveys, Change Orders, designs or specifications, or (2) the giving of or the failure to give directions or instructions by the Architect, his agents or employees provided such giving or failure to give is the primary cause of the injury or damage.

## ARTICLE 12
### SUBCONTRACTS

**12.1** A Subcontractor is a person who has a direct contract with the Contractor to perform any of the Work at the site.

**12.2** Unless otherwise specified in the Contract Docu-

ments or in the Instructions to Bidders, the Contractor, as soon as practicable after the award of the Contract, shall furnish to the Architect in writing a list of the names of Subcontractors proposed for the principal portions of the Work. The Contractor shall not employ any Subcontractor to whom the Architect or the Owner may have a reasonable objection. The Contractor shall not be required to employ any Subcontractor to whom he has a reasonable objection. Contracts between the Contractor and the Subcontractor shall be in accordance with the terms of this Agreement and shall include the General Conditions of this Agreement insofar as applicable.

## ARTICLE 13
### SEPARATE CONTRACTS

The Owner has the right to let other contracts in connection with the Work and the Contractor shall properly cooperate with any such other contractors.

## ARTICLE 14
### ROYALTIES AND PATENTS

The Contractor shall pay all royalties and license fees. The Contractor shall defend all suits or claims for infringement of any patent rights and shall save the Owner harmless from loss on account thereof.

## ARTICLE 15
### ARBITRATION

All claims or disputes arising out of this Contract or the breach thereof shall be decided by arbitration in accordance with the Construction Industry Arbitration Rules of the American Arbitration Association then obtaining unless the parties mutually agree otherwise. Notice of the demand for arbitration shall be filed in writing with the other party to the Contract and with the American Arbitration Association and shall be made within a reasonable time after the dispute has arisen.

## ARTICLE 16
### TIME

**16.1** All time limits stated in the Contract Documents are of the essence of the Contract.

**16.2** If the Contractor is delayed at any time in the progress of the Work by changes ordered in the Work, by labor disputes, fire, unusual delay in transportation, unavoidable casualties, causes beyond the Contractor's control, or by any cause which the Architect may determine justifies the delay, then the Contract Time shall be extended by Change Order for such reasonable time as the Architect may determine.

## ARTICLE 17
### PAYMENTS

**17.1** Payments shall be made as provided in Article 5 of this Agreement.

**17.2** Payments may be withheld on account of (1) defective Work not remedied, (2) claims filed, (3) failure of the Contractor to make payments properly to Sub-

**AIA DOCUMENT A107** · SMALL CONSTRUCTION CONTRACT · SEPTEMBER 1970 EDITION · AIA®
©1970 · THE AMERICAN INSTITUTE OF ARCHITECTS, 1735 NEW YORK AVE., N.W., WASHINGTON, D.C. 20006

contractors or for labor, materials, or equipment, (4) damage to another contractor, or (5) unsatisfactory prosecution of the Work by the Contractor.

**17.3** Final payment shall not be due until the Contractor has delivered to the Owner a complete release of all liens arising out of this Contract or receipts in full covering all labor, materials and equipment for which a lien could be filed, or a bond satisfactory to the Owner indemnifying him against any lien.

**17.4** The making of final payment shall constitute a waiver of all claims by the Owner except those arising from (1) unsettled liens, (2) faulty or defective Work appearing after Substantial Completion, (3) failure of the Work to comply with the requirements of the Contract Documents, or (4) terms of any special guarantees required by the Contract Documents. The acceptance of final payment shall constitute a waiver of all claims by the Contractor except those previously made in writing and still unsettled.

## ARTICLE 18
### PROTECTION OF PERSONS AND PROPERTY

The Contractor shall be responsible for initiating, maintaining, and supervising all safety precautions and programs in connection with the Work. He shall take all reasonable precautions for the safety of, and shall provide all reasonable protection to prevent damage, injury or loss to (1) all employees on the Work and other persons who may be affected thereby, (2) all the Work and all materials and equipment to be incorporated therein, and (3) other property at the site or adjacent thereto. He shall comply with all applicable laws, ordinances, rules, regulations and orders of any public authority having jurisdiction for the safety of persons or property or to protect them from damage, injury or loss. All damage or loss to any property caused in whole or in part by the Contractor, any Subcontractor, any Subsubcontractor or anyone directly or indirectly employed by any of them, or by anyone for whose acts any of them may be liable, shall be remedied by the Contractor, except damage or loss attributable to faulty Drawings or Specifications or to the acts or omissions of the Owner or Architect or anyone employed by either of them or for whose acts either of them may be liable but which are not attributable to the fault or negligence of the Contractor.

## ARTICLE 19
### CONTRACTOR'S LIABILITY INSURANCE

The Contractor shall purchase and maintain such insurance as will protect him from claims under workmen's compensation acts and other employee benefit acts, from claims for damages because of bodily injury, including death, and from claims for damages to property which may arise out of or result from the Contractor's operations under this Contract, whether such operations be by himself or by any Subcontractor or anyone directly or indirectly employed by any of them. This insurance shall be written for not less than any limits of liability specified as part of this Contract, or required by law, whichever is the greater, and shall include contractual liability insurance as applicable to the Contractor's obli-

gations under Paragraph 11.10. Certificates of such insurance shall be filed with the Owner.

## ARTICLE 20
### OWNER'S LIABILITY INSURANCE

The Owner shall be responsible for purchasing and maintaining his own liability insurance and, at his option, may maintain such insurance as will protect him against claims which may arise from operations under the Contract.

## ARTICLE 21
### PROPERTY INSURANCE

**21.1** Unless otherwise provided, the Owner shall purchase and maintain property insurance upon the entire Work at the site to the full insurable value thereof. This insurance shall include the interests of the Owner, the Contractor, Subcontractors and Sub-subcontractors in the Work and shall insure against the perils of Fire, Extended Coverage, Vandalism and Malicious Mischief.

**21.2** Any insured loss is to be adjusted with the Owner and made payable to the Owner as trustee for the insureds, as their interests may appear, subject to the requirements of any mortgagee clause.

**21.3** The Owner shall file a copy of all policies with the Contractor prior to the commencement of the Work.

**21.4** The Owner and Contractor waive all rights against each other for damages caused by fire or other perils to the extent covered by insurance provided under this paragraph. The Contractor shall require similar waivers by Subcontractors and Sub-subcontractors.

## ARTICLE 22
### CHANGES IN THE WORK

**22.1** The Owner without invalidating the Contract may order Changes in the Work consisting of additions, deletions, or modifications, the Contract Sum and the Contract Time being adjusted accordingly. All such Changes in the Work shall be authorized by written Change Order signed by the Owner or the Architect as his duly authorized agent.

**22.2** The Contract Sum and the Contract Time may be changed only by Change Order.

**22.3** The cost or credit to the Owner from a Change in the Work shall be determined by mutual agreement.

## ARTICLE 23
### CORRECTION OF WORK

The Contractor shall correct any Work that fails to conform to the requirements of the Contract Documents where such failure to conform appears during the progress of the Work, and shall remedy any defects due to faulty materials, equipment or workmanship which appear within a period of one year from the Date of Substantial Completion of the Contract or within such longer period of time as may be prescribed by law or by the terms of any applicable special guarantee required by the Contract Documents. The provisions of this Article 23 apply to Work done by Subcontractors as well as to Work done by direct employees of the Contractor.

**AIA DOCUMENT A107** • SMALL CONSTRUCTION CONTRACT • SEPTEMBER 1970 EDITION • AIA®
©1970 • THE AMERICAN INSTITUTE OF ARCHITECTS, 1735 NEW YORK AVE., N.W., WASHINGTON, D.C. 20006   **6**

## ARTICLE 24
### TERMINATION BY THE CONTRACTOR

If the Architect fails to issue a Certificate of Payment for a period of thirty days through no fault of the Contractor, or if the Owner fails to make payment thereon for a period of thirty days, the Contractor may, upon seven days' written notice to the Owner and the Architect, terminate the Contract and recover from the Owner payment for all Work executed and for any proven loss sustained upon any materials, equipment, tools, and construction equipment and machinery, including reasonable profit and damages.

## ARTICLE 25
### TERMINATION BY THE OWNER

If the Contractor defaults or neglects to carry out the Work in accordance with the Contract Documents or fails to perform any provision of the Contract, the Owner may, after seven days' written notice to the Contractor and without prejudice to any other remedy he may have, make good such deficiencies and may deduct the cost thereof from the payment then or thereafter due the Contractor or, at his option, may terminate the Contract and take possession of the site and of all materials, equipment, tools, and construction equipment and machinery thereon owned by the Contractor and may finish the Work by whatever method he may deem expedient, and if the unpaid balance of the Contract Sum exceeds the expense of finishing the Work, such excess shall be paid to the Contractor, but if such expense exceeds such unpaid balance, the Contractor shall pay the difference to the Owner.

**AIA DOCUMENT A107** • SMALL CONSTRUCTION CONTRACT • SEPTEMBER 1970 EDITION • AIA®
©1970 • THE AMERICAN INSTITUTE OF ARCHITECTS, 1735 NEW YORK AVE., N.W., WASHINGTON, D.C. 20006

This Agreement executed the day and year first written above.

OWNER                                                   CONTRACTOR

**AIA DOCUMENT A107** • SMALL CONSTRUCTION CONTRACT • SEPTEMBER 1970 EDITION • AIA®
©1970 • THE AMERICAN INSTITUTE OF ARCHITECTS, 1735 NEW YORK AVE., N.W., WASHINGTON, D.C. 20006       **8**

"This document has been reproduced with the permission of The American Institute of Architects. Further reproduction is not authorized".

Owner of Project:

Address:

Re: Name of Construction Project

Gentlemen:

Your instructions are requested concerning the insurance provisions to be incorporated in the Supplementary Conditions of the Contract for the proposed Construction Project.

The Contract Specifications will require the Contractor to carry certain forms of insurance such as Workmen's Compensation, Public Liability, Property Damage and Automobile Liability. For your convenience Enclosure "A" may be used to instruct us regarding the limits and kinds of coverage which you wish specified.

There are certain other insurance coverages which are usually the responsibility of the Owner, such as: Builder's Risk Fire and Extended Coverage or All Physical Loss; Owner's Contingent Liability Insurance, Boiler Coverage, etc. Your instructions to us in connection with these coverages can be indicated on Enclosure "B."

We recommend that you consult your insurance counsellor concerning the limits and kinds of insurance that should be provided to insure the Project adequately. We shall be glad to supply any additional information which he may require. Please advise us as soon as possible in order that we may complete the Contract Documents.

Very truly yours,

_____

Architect

**12** CHAPTER 7 • INSURANCE AND BONDS OF SURETYSHIP • ARCHITECT'S HANDBOOK OF PROFESSIONAL PRACTICE • JAN. 1969 ED. • AIA® • ©1969 • THE AMERICAN INSTITUTE OF ARCHITECTS • 1735 NEW YORK AVENUE, N.W., WASHINGTON, D. C. 20006

ENCLOSURE "A"

TO _____  DATE:
   *(Architect)*

FROM _____  SUBJECT: Contractor's Insurance for.
   *(Owner)*                                      (Name of Construction Project)

Gentlemen:

This will acknowledge your request for instructions regarding amounts and kinds of insurance to be specified in the Supplementary Conditions. The Contractor and all Subcontractors, at their own expense, shall provide and maintain insurance as follows, in companies acceptable to the Owner.

1. **Workmen's Compensation** as required by all applicable Federal, State, Maritime or other laws including Employers Liability with a limit of at least:                                          $_____
   *(indicate limit)*

2. **Comprehensive General Liability** including Contractors Liability; Contingent Liability; Contractual Liability; Completed Operations and Products Liability all on the occurrence basis with Personal Injury Coverage and broad form Property Damage. Remove the XCU exclusion relating to Explosion, Collapse and Underground Property Damage. Completed Operations Liability shall be kept in force for at least two years after the date of final completion.
   *(indicate limit)*

   **Personal Injury**
   Each Person                                                                          $_____
   Each Occurrence                                                                      $_____

   **Property Damage**
   Each Accident                                                                        $_____
   Aggregate                                                                            $_____

3. **Comprehensive Automobile Liability** including non-ownership and hired car coverage as well as owned vehicles:
   *(indicate limit)*

   **Bodily Injury**
   Each Person                                                                          $_____
   Each Occurrence                                                                      $_____

   **Property Damage**
   Each Occurrence                                                                      $_____

4. **Bonds or Other Insurance** which we require the Contractor to provide:
   *Kind*                                                                               *Amount*

Performance Bond, AIA Document A311
_____

Labor and Material Payment Bond, AIA Document A311
_____

_____

_____

The Contractor shall furnish the Owner with satisfactory evidence of the required insurance and/or Bonds with a provision that at least fifteen days prior written notice will be given to the Owner in the event of cancellation or material change.

Very truly yours,

_____
Owner
_____

CHAPTER 7 • INSURANCE AND BONDS OF SURETYSHIP • ARCHITECT'S HANDBOOK OF PROFESSIONAL PRACTICE • JAN. 1969 ED. • AIA® • ©1969 • THE AMERICAN INSTITUTE OF ARCHITECTS • 1735 NEW YORK AVENUE, N.W., WASHINGTON, D. C. 20006   **13**

# CERTIFICATE OF INSURANCE
*AIA Document G705*

This certifies to the Addressee shown below that the following described policies, subject to their terms, conditions and exclusions, have been issued to:

NAME AND ADDRESS OF INSURED

COVERING  (SHOW PROJECT NAME AND/OR NUMBER AND LOCATION)

Addressee:

```
┌                              ┐



└                              ┘
```

Date

| KIND OF INSURANCE | POLICY NUMBER | Inception Date | Expiration Date | | LIMITS OF LIABILITY (in thousands of dollars) |
|---|---|---|---|---|---|
| 1. (a) Workmen's Comp. | | | | $ ///////// | Statutory Workmen's Compensation |
| (b) Employers' Liability | | | | $ | One Accident and Aggregate Disease |
| 2. Comprehensive General Liability | | | | $ | Each Person — Premises and Operations |
| | | | | $ | Each Person — Elevators |
| | | | | $ | Each Person — Independent Contractors |
| (a) Bodily Injury | | | | $ | Each Person — PRODUCTS INCLUDING COMPLETED OPERATIONS |
| Including | | | | $ | Each Person — Contractual |
| Personal Injury | | | | $ | Each Occurrence — |
| | | | | $ | Aggregate — PRODUCTS INCLUDING COMPLETED OPERATIONS |
| | | | | $ | Each Occurrence — Premises — Operations |
| | | | | $ | Each Occurrence — Elevators |
| | | | | $ | Each Occurrence — INDEPENDENT CONTRACTOR |
| (b) Property Damage | | | | $ | Each Occurrence — PRODUCTS INCLUDING COMPLETED OPERATIONS |
| | | | | $ | Each Occurrence — Contractual |
| | | | | $ | Aggregate — |
| | | | | $ | Aggregate — OPERATIONS PROTECTIVE PRODUCTS AND CONTRACTUAL |
| 3. Comprehensive Automobile Liability | | | | $ | Each Person — |
| (a) Bodily Injury | | | | $ | Each Occurrence — |
| (b) Property Damage | | | | $ | Each Accident — |
| 4. (Other) | | | | | |

**UNDER GENERAL LIABILITY POLICY OR POLICIES**                                              Yes          No
1.  Does Property Damage Liability Insurance shown include coverage for XC and U hazards? . . .  _____   _____
2.  Is Occurrence Basis Coverage provided under Property Damage Liability? . . . . . . . . .  _____   _____
3.  Is Broad Form Property Damage Coverage provided for this Project? . . . . . . . . . . .  _____   _____
4.  Does Personal Injury Liability Insurance include coverage for personal injury sustained by any person as a result of an offense directly or indirectly related to the employment of such person by the Insured? . . . . . . . . . . . . . . . . . . . . . . . . . . . . . .  _____   _____
5.  Is coverage provided for Contractual Liability (including indemnification provision) assumed by Insured? . . . . . . . . . . . . . . . . . . . . . . . . . . . . . . . . .  _____   _____

**UNDER AUTOMOBILE LIABILITY POLICY OR POLICIES**
1.  Does coverage shown above apply to non-owned and hired automobiles? . . . . . . . . . .  _____   _____
2.  Is Occurrence Basis Coverage provided under Property Damage Liability? . . . . . . . . .  _____   _____

**CANCELLATION**

In the event of cancellation of any of the foregoing, fifteen (15) days written notice shall be given to the party to whom this Certificate is addressed.

NAME OF INSURANCE COMPANY

ADDRESS

SIGNATURE OF AUTHORIZED REPRESENTATIVE

CHAPTER 7 • INSURANCE AND BONDS OF SURETYSHIP • ARCHITECT'S HANDBOOK OF PROFESSIONAL PRACTICE • JAN. 1969 ED. • AIA® • ©1969 • THE AMERICAN INSTITUTE OF ARCHITECTS • 1735 NEW YORK AVENUE, N.W., WASHINGTON, D. C. 20006

16

## PROTOTYPE LETTER 3 • OWNER'S INSTRUCTION REGARDING OWNER'S INSURANCE

ENCLOSURE "B"

TO _____ DATE:
  (Architect)

FROM _____ SUBJECT: Owner's Insurance for
  (Owner)                                              (Name of Construction Project)

Gentlemen:

You are hereby instructed to indicate in the Specifications that the Owner will provide the following kinds of insurance for the above Project:

1. **Owner's Contingent Liability**

   **Personal Injury**

   *(indicate limit)*

   Each Person                                                          $_____
   Each Occurrence                                                      $_____

   **Property Damage**
   Each Occurrence                                                      $_____
   Aggregate                                                            $_____

2. **Builder's Risk**—Fire, Extended Coverage and Vandalism Insurance on the completed value form for the full insurable value of the Work in the names of the Owner, the Contractor and all Subcontractors as their interests may appear (Subparagraph 11.3.1 of the General Conditions).

3. **Steam Boiler and Machinery** (Subparagraph 11.3.2 of the General Conditions)
   a. Limit                                                             $_____
   b. Objects to be insured  *(list objects)*

4. **Other Insurance**
   (Special coverage, etc.)              Kind                           Amount

   _____                     $_____

   _____                     $_____

   _____                     $_____

Very truly yours,


_____
Owner

# 12

## General Requirements

The General Requirements of the specifications consist of certain sections listed under Division 1 of the *Uniform Construction Index* (see Chapter 19). The sections recommended by the *Uniform Construction Index* for inclusion under this division are the following:

| Section No. | Title |
|---|---|
| 01010 | Summary of Work |
| 01100 | Alternatives |
| 01200 | Project Meetings |
| 01300 | Submittals |
| 01400 | Quality Control |
| 01500 | Temporary Facilities and Controls |
| 01600 | Material and Equipment |
| 01700 | Project Closeout |

Since the General Requirements and the proposed sections included therein are a relatively recent innovation (1972), they have not yet been exposed to widespread use and experience. The recommendations originally suggested under the CSI Format, which were promulgated in July 1964 for inclusion under the General Requirements, consisted of alternates, alterations, inspections, tests, allowances, and temporary facilities. From the standpoint of time and usage the value of some of the sections proposed under the *Uniform Construction Index* is yet to be determined.

The original purpose of the General Requirements in the first CSI Format, issued in 1964, was to provide a place for the nonlegal, nontechnical requirements that are required by the contractor in order to construct the project. All those requirements of a general nature not suitable for inclusion under the technical sections were also expected to be set forth here by the early proponents of the CSI Format. As stated in Chapter 11, the supplementary conditions had become the catch-all section in which were specified temporary utilities, temporary facilities, and a host of other requirements not of a legal nature. It was the intention that the establishment of a division entitled "General Requirements" would create a convenient place for instructions to the contractor that could not logically be placed anywhere else.

The *Uniform Construction Index* does not list alterations anywhere. There is, however, a need to describe this work, which occurs quite frequently. Since the nature of alterations involves many trades, and since it is preferable to write a performance rather than a descriptive type of specification for this work, a general section under Division 1 places the responsibility directly on the contractor to coordinate the various activities of the subcontractors involved in this operation.

Most drawings for alterations simply indicate existing partitions to be removed or an existing door or window to be removed and blocked in, without any additional details. The specifications, too, for this work usually refer to matching existing conditions so that the work of alterations cannot usually be described specifically, but must be covered generally.

A specification for alterations, in general performance-type language, that has been used quite successfully is included with the sample illustrations which follow. Sample sections delineating the scope and content of Division 1 are appended herewith. *Note*: The designation *NTS* used in the illustrations means note to specifier.

SECTION 01010
SUMMARY OF WORK

*NTS:* Par. 1. Describe briefly the work to be performed under this contract.

1. WORK UNDER THIS CONTRACT

*NTS:* Par. 2.    If work is to be performed under multiple contracts, that is, HVAC, Plumbing, and Electrical, enumerate these contracts.

2. WORK PERFORMED UNDER SEPARATE CONTRACTS

*NTS:* Par. 3.    This should be confined to work shown as N.I.C., such as carpeting, draperies, and so on.

3. WORK BY OTHERS

*NTS:* Par. 4.    List items furnished by owner. Delineate who installs.

4. ITEMS FURNISHED BY OWNER

   a. The following items are furnished by the Owner and installed by the Contractor:

   b. The following items are furnished and installed by the Owner.

5. ALLOWANCES

   a. *General.*    See General Conditions, Par. 4.8.1.

   *NTS:* Par. 5b.    Make a tabulation of the allowances specified.

   b. *Schedule of Allowances*

   1. ITEM          SECTION

   *NTS:* Par. 6.    Use as necessary.

6. WORK TO BE PERFORMED LATER

   *NTS:* Par. 7.    Use primarily for multiple contracts and specify requirements.

7. COORDINATION

   *NTS:* Par. 8.    Use only as an extension of General Conditions where warranted when separate prime contracts are involved.

8. CUTTING AND PATCHING

9. APPLICABLE CODES

   a. All references to codes, specifications, and standards referred to in the Specification Sections and on the Drawings shall mean, and are intended to be, the lastest edition, amendment and/or revision of such reference standard in effect as of the date of these Contract Documents.

10. ABBREVIATIONS AND SYMBOLS

   a. Reference to a technical society, institution, association, or governmental authority is made in the Specifications in accordance with the following abbreviations:

| | |
|---|---|
| AA | Aluminum Association |
| AAMA | Architectural Aluminum Manufacturers Association |
| AASHO | American Association of State Highway Officials |
| ACI | American Concrete Institute |
| AFI | Air Filter Institute |
| AGA | American Gas Association |
| AGC | Associated General Contractors of America |
| AGMA | American Gear Manufacturers Association |
| AIA | American Institute of Architects |
| AIA | American Insurance Association (Formerly NBFU) |
| AIEE | American Institute of Electrical Engineers |
| AISC | American Institute of Steel Construction |
| AISI | American Iron and Steel Institute |
| ALS | American Lumber Standards |
| AMA | Acoustical Materials Association |
| AMCA | Air Moving and Conditioning Association |
| ANSI | American National Standards Institute (Formerly USAS) |
| APA | American Plywood Association |
| ARI | Air Conditioning and Refrigeration Institute |
| ASHRAE | American Society of Heating, Refrigerating, and Air Conditioning Engineers |
| ASME | American Society of Mechanical Engineers |
| ASTM | American Society for Testing and Materials |
| AWI | Architectural Woodwork Institute |
| AWPA | American Wood Preservers' Association |
| AWPI | American Wood Preservers' Institute |
| AWS | American Welding Society |
| AWWA | American Water Works Association |
| CABRA | Copper and Brass Research Association |
| CS | Commercial Standard, U.S. Department of Commerce |
| CSI | Construction Specification Institute |
| FGJA | Flat Glass Jobbers Association |
| FIA | Factory Insurance Association |
| FM | Factory Mutual |
| FS | Federal Specification |

| | |
|---|---|
| GA | Gypsum Association |
| IEEE | Institute of Electric and Electronics Engineers |
| IES | Illuminating Engineering Society |
| MIA | Marble Institute of America |
| MLMA | Metal Lath Manufacturers Association |
| MS | Military Specification |
| MSTD | Military Standard |
| NAAMM | National Association of Architectural Metal Manufacturers |
| NAFM | National Association of Fan Manufacturers |
| NBFU | National Board of Fire Underwriters (currently American Insurance Association) |
| NBS | National Bureau of Standards |
| NEC | National Electric Code |
| NEMA | National Electrical Manufacturers Association |
| NFC | National Fire Code |
| NFPA | National Electrical Protection Association |
| NHLA | National Hardwood Lumber Association |
| NLMA | National Lumber Manufacturers Association |
| NTMA | National Terrazzo and Mosaic Association |
| NWMA | National Woodwork Manufacturers Association |
| PEI | Porcelain Enamel Institute |
| RTI | Resilient Tile Institute |
| SBI | Steel Boiler Institute |
| SCPI | Structural Clay Products Institute |
| SDI | Steel Deck Institute |
| SJI | Steel Joist Institute |
| SMACNA | Sheet Metal and Air Conditioning Contractors National Association |
| SPR | Simplified Practice Recommendation, U.S. Department of Commerce |
| SSPC | Steel Structures Painting Council |
| TCA | Tile Council of America |
| UL | Underwriters' Laboratories |

11. COLOR SCHEDULE

a. A color schedule will be issued by the architect. The Contractor, his subcontractors, and material suppliers shall cooperate in furnishing required color samples to aid in the final selections. Where special colors are selected by the Architect, furnish accurate reproductions of these colors, on actual material to be furnished to the project, for review.

## SECTION 01100
## ALTERNATIVES

1. SCOPE
   a. Perform work required for complete execution of accepted alternatives. Amount of alternative prices shall include cost of modifications made necessary including overhead and profit.
   b. Work for alternatives shall comply with applicable provisions of the Contract Documents, except as otherwise specified herein.
2. LIST OF ALTERNATIVES
   Alternative No. 1    Vinyl Asbestos Tile
   a. State the amount to be deducted from the Base Bid if vinyl asbestos tile is used in all corridors in lieu of terrazzo as indicated in Finish Schedule.
   b. Vinyl asbestos tile shall conform to FS SS-T-312, 9 in. × 9 in. × 1/8 in., manufacturer's standard color as selected.
   c. Install in accordance with the requirements of the Resilient Tile Association.
   Alternative No. 2    Stainless Steel Elevator Entrances
   a. State the amount to be added to the Base Bid is stainless steel is used for elevator entrances in lieu of baked enamel hollow metal.
   b. Stainless steel shall be type 302, 14 U.S. gage, with a No. 4 finish.
   Alternative No. 3    Emergency Generator
   a. State the amount to be added to the Base Bid if Emergency Generator as specified in Section 16000 is furnished and installed.

## SECTION 01200
## PROJECT MEETINGS

*NTS:* Par. 1.    Where required, specify requirements for special preconstruction conferences.

1. PRECONSTRUCTION CONFERENCES
2. PROGRESS MEETINGS
   a. The Contractor and any subcontractors, material men, or vendors whose presence is necessary or requested must attend meetings (referred to as progress meetings) when called by the

Architect or his representative for the purpose of discussing the execution of the work. Each of such meetings will be held at the time and place designated by Architect or his representative. All decisions, instructions, and interpretations given by the Architect or his representative at these meetings shall be binding and conclusive on the Contractor. The proceedings of these meetings will be recorded and the Contractor will be furnished a reasonable number of copies for his use and for distribution to the various subcontractors, material men, and vendors involved.

*NTS:* Use Par. 2b for multiple contracts.

b. Progress meetings may also be called by the General Construction Contractor for the purpose of coordinating, expediting, and scheduling the work of all contracts. Other contractors and their subcontractors, material men, or vendors whose presence is necessary or requested are required to attend.

## SECTION 01300
## SUBMITTALS

*NTS:* Par. 1.  If further delineation of a construction schedule is required, amplify this paragraph.

1. CONSTRUCTION SCHEDULES

a. See General Conditions, Par. 4.11, for submission of a progress schedule.

*NTS:* Par. 2.  If a CPM is required, specify provisions here for submission of report.

2. NETWORK ANALYSIS

*NTS:* Par. 3.  Specify requirements here if required.

3. PROGRESS REPORTS
4. SURVEY DATA

a. Be responsible for properly laying out the work, and for lines and measurements for the work executed under the Contract Documents. Verify the figures shown on the Drawings before laying out the work, and report errors or inaccuracies in writing to the Architect before commencing work. The Architect or his representative will in no case assume the responsibility for laying out the work.

b. Establish necessary reference lines and permanent bench marks from which building lines and elevations shall be established.

Engage a registered land surveyor for this purpose to lay out the work. Establish not less than two such bench marks in widely separated locations. Be responsible for the proper location and level of the work and for the maintenance of the reference lines and bench marks. Establish bench marks and axis lines at each floor showing exact floor elevations and other lines and dimensional reference points as required for the information and guidance of all trades; field checking of the structure and surveys thereof as may be required by the technical sections of the specifications; the marking and layout of walls and partitions; and the taking of settlement readings as hereinafter specified.

c. Take settlement readings of the work at a predetermined number of points selected by the Architect. Take readings weekly until the work is substantially completed or until such time as the Architect directs. Record survey data and submit to the Architect.

d. The mechanical and electrical trades shall be responsible for the layout of the duct work, piping, and conduit, based on the reference lines and bench marks established.

e. Furnish certification from a registered land surveyor who shall verify periodically that portions of the work are located in accordance with the Drawings and at the required elevations. Upon completion of foundation walls, prepare and submit to the Architect a certificate survey showing that dimensions, elevations, angles and the location of the building is in accordance with the Contract Documents. When enclosing walls are completed, a further survey shall be submitted, certifying their location and plumbness.

5. SHOP DRAWINGS AND SAMPLES (SUBMITTALS AND DISTRIBUTION)

a. *General.* The contractual requirements for shop drawings and samples are specified in the General Conditions. Contractor shall submit shop drawings and samples accompanied by the Architect's "Shop Drawing and Sample Transmittals" form.

*NTS:* Select subparagraph 1 or 2.

1. The Architect's transmittal forms shall be furnished by the Contractor.

2. The Architect will furnish the transmittal forms to the Contractor for his use.

b. *Preparation of Submittal Form.*   Fill out transmittal form.
c. *Resubmissions.*   Resubmittal procedure shall follow the same procedure as the initial submittal.
d. *Architect's Action on Transmittal Form*
   1. Incomplete or erroneous transmittals will be returned without action.
   2. The remainder of the transmittal will be filled out by Architect.
e. *Submittal Procedures by Contractor*
   1. Architectural items as follows:
      (a) One sepia and one print of shop drawings with transmittal forms.
      (b) Three samples with transmittal forms.
      (c) Six copies of brochures with transmittal forms.
f. *Architect's Review*
   1. Architect will process the submission and indicate the appropriate action on the submission and the transmittal.
6. CONSTRUCTION PHOTOGRAPHS
   a. During the progress of the work have black and white photographs taken once a month, consisting of three views, all taken where directed by the Architect. The prints shall be 8 in. × 10 in., linen backed. At the completion of all work, five final photographs shall be taken as directed by the Architect.
   b. One print of each photograph shall be mailed to the Owner and two prints to the Architect. The photographs shall be neatly labeled, dated, and identified in a title box in the lower right-hand corner, showing the date of exposure, project name and location, and direction of view.
   c. All negatives shall be retained by the photographer until completion of the work, at which time they shall become the property of the Owner. Obtain and deliver the negatives to the Owner.

SECTION 01400
QUALITY CONTROL

1. GENERAL
   a. Provide and maintain an effective Contractor Quality Control (CQC) program and perform sufficient inspections and tests of all items of work, including those of subcontractors, to

ensure compliance with Contract Documents. Include surveillance and tests specified in the technical sections of the Specifications. Furnish appropriate facilities, instruments, and testing devices required for performance of the quality control function. Controls must be adequate to cover construction operations and be keyed to the construction sequence.

2. CONTROL OF ON-SITE CONSTRUCTION
   a. Include a control system for the following phases of inspection:
      1. *Preparatory Inspection.*   Perform this inspection prior to beginning work on any definable feature of work. Include a review of contract requirements with the supervisors directly responsible for the performance of the work; check to assure that materials, products, and equipment have been tested, submitted, and approved; check to assure that provisions have been made for required control testing; examine the work area to ascertain that preliminary work has been completed; physically examine materials and equipment to assure that they conform to shop drawings and data and that the materials and equipment are on hand.
      2. *Initial Inspection.*   Perform this inspection as soon as work commences on a representative portion of a particular feature of workmanship; review control testing for compliance with contract requirements.
      3. *Follow-up Inspections.*   Perform these inspections on a regular basis to assure continuing compliance with contract requirements until completion of that particular work.
      4. *Documentation of CQC Report.*   Identify the inspections hereinbefore specified and document in the CQC report with a brief description of the subject matter covered and the personnel involved.

3. CONTROL OF OFF-SITE OPERATIONS
   a. Perform factory quality control inspection for items fabricated or assembled off-site as opposed to "off-the-shelf" items. The CQC representative at the fabricating plant shall be responsible for release of the fabricated items for shipment to the job site. The CQC representative at the job site shall receive the item and note any damage incurred during shipment. The Contractor shall be responsible for protecting and maintaining the item in good condition throughout the period of on-site

storage and during erection or installation. Although any item found to be faulty may be rejected before it is used, final acceptance of an item by the owner is based on its satisfactory incorporation into the work and acceptance of the completed project.

4. QUALITY CONTROL ORGANIZATION

   a. The CQC staff shall function completely independently of the Contractor's job-supervisory staff and shall be composed of the following members:

      1. *Architect.* Minimum five years' experience in supervising and inspecting equivalent construction. He shall be in charge of the quality control staff and program during the life of the project and inspect all contract operations.

      2. *Electrical Engineer.* Minimum five years' experience in installing, testing, and operating equipment and services equivalent to that required on this project. He shall prepare, supervise, monitor, and record all data on electrical, electronic, and other types of tests required on this project.

      3. *Mechanical Engineer.* Minimum five years' experience in installing, testing, and operating equipment and services equivalent to that required on this project. He shall inspect all mechanical equipment and systems required during the life of the project.

      4. *Mechanical Engineering Technician.* Minimum three years' experience in inspecting and testing installation of mechanical equipment and systems similar to those required on this project. He shall be on duty at all times during the installation and testing of elevators, plumbing systems, refrigeration, and air conditioning systems under this contract. He shall personally supervise and certify to the correctness of balancing of water, balancing of heating system, balancing of air conditioning systems, and performance testing of mechanical installations. He shall be on duty after completion of installation and through final acceptance and inspection tests on the balancing operation and performance of all these systems.

      5. *Electrical Engineering Technician.* Minimum three years' experience in inspection of electrical equipment and systems similar to those required on this project. He shall be

on duty at all times during the installation and testing of radio and television systems until they are finally accepted.

6. *Civil Engineer.*　Minimum three years' experience in the inspection and testing of concrete and precast concrete equivalent to that required on this project. He shall be on duty at all times during the production of concrete and precast concrete to supervise and inspect concrete operations and perform the following:

   (a) Sampling and testing of concrete materials and concrete.
   (b) Storage of materials.
   (c) Batching.
   (d) Placement.
   (e) Formwork.
   (f) Curing.

7. *Technicians.*　Provide specialists to ensure capability of complying with CQC as specified.

## 5. TESTING LABORATORY

a. The Contractor may arrange for a testing laboratory to provide on-site services required in lieu of direct employment of personnel. The tests shall be documented and certified. All compliance inspections shall record both conforming and defective items with an explanation for the cause of rejection, proposed remedial action, and corrective action taken. Individual daily reports will be required of each inspector and technician covering the feature which each is assigned and a consolidated daily report signed by the person in charge of the quality control staff.

## 6. SCHEDULE OF CQC PLAN

a. Furnish a schedule outlining the procedures, instructions, and reports to be used as follows:

1. Quality control organization.
2. Qualifications of personnel.
3. Authority and responsibility of personnel.
4. Schedule of inspection personnel.
5. Test methods.
6. Methods of performing and documenting quality control operations.

## 7. REPORTS

a. Inspection shall be recorded and submitted daily on approved forms certifying items correctly installed and items found to be defective with a statement on corrective measures taken.

8. TESTING AND INSPECTION DEVICES
   a. Either the Contractor or his testing laboratory shall provide and maintain all measuring and testing devices, laboratory equipment, instruments, and supplies necessary to accomplish the required testing and inspection. All measuring and testing devices shall be calibrated periodically against certified standards.

9. TESTING AND INSPECTION REQUIREMENTS
   a. Where technical sections of the specifications require inspection and testing by a testing laboratory, engage a reputable, recognized testing laboratory, experienced in the type of work to be perfomed. The representative of the testing laboratory shall be on the work site as necessary for sampling, inspection, and testing in accordance with the contract provisions. Submit written reports of results within three days after completion of tests.

10. LATEST DOCUMENTS
    a. The Contractors Quality Control system shall provide for procedures which will assure that the latest Contract Documents, shop drawings, and instructions required by the Contract are used for fabrication, testing, and inspection.

SECTION 01500
TEMPORARY FACILITIES AND CONTROLS

1. GENERAL
   a. Arrange for and provide temporary facilities and controls as specified herein and as required for the proper and expeditious prosecution of the work. Pay all costs, except as otherwise specified, until final acceptance of the work unless the owner makes arrangements for the use of completed portions of the work after substantial completion in accordance with the provisions of the General Conditions.
   b. Make all temporary connections to utilities and services in locations acceptable to the Owner, Architect, and local authorities having jurisdiction thereof; furnish all necessary labor and materials, and make all installations in a manner subject to the acceptance of such authorities and the Architect; main-

tain such connections; remove the temporary installation and connections when no longer required; restore the services and sources of supply to proper operating condition.

c. Pay all costs for temporary electrical power, temporary water, and temporary heating.

## 2. PROJECT IDENTIFICATION

a. No signs or advertisements will be allowed to be displayed on the premises without the approval of the Architect.

b. One construction sign shall be provided by the Contractor and shall be subject to the review of the Architect and the approval of the Owner. Text and lettering shall be provided for at a later date.

c. Erect the construction sign on the site where directed approximately 4 ft × 8 ft in size, of 3/4 in. plywood with suitable frame, moldings, and supports. Use Douglas Fir Overlaid Plywood, Grade B-B high density, exterior, good two sides, complying with PS-1. The sign shall be primed and given two coats of approved white paint. Lettering shall be black of an approved type, size, and layout as directed by the Architect. Sign shall contain the name of the building, Owner, Architect, Contractor, and such other reasonable information as the Architect or Owner may require.

## 3. MATERIAL HOIST

a. Provide a material hoist as required for normal use by all trades. Provide all necessary guards, signals, safety devices, and so on, required for safe operations, and suitable runways from the hoists to each floor level and roof. The construction and operation of the material hoist shall comply with all applicable requirements of ANSI A10.5, the AGC *Manual of Accident Prevention in Construction*, and to all applicable state and municipal codes. Prohibit the use of the material hoist for transporting personnel.

b. Special rigging and hoisting facilities shall be provided by each trade requiring same.

## 4. RODENT CONTROL

a. Institute an effective program of rodent control for the entire site within the construction limits. Cooperate with local authorities and provide the regular services of an experienced exterminator who shall visit the site at least once a month for the entire construction period. Provide marked metal containers for all edible rubbish and enforce their use by all employees. Containers shall be emptied and the contents removed from the site as often as required to maintain an adequate rodent control program. If the program of rodent control utilized is not effective, take whatever steps are necessary to rid the project of rodents, and such action shall not be the basis of a claim for additional compensation or damages.

## 5. TEMPORARY CONSTRUCTION OPENINGS

a. Provide openings in slabs, walls, and partitions where required for moving in large pieces of equipment of all types. Close and/or restore all openings and finish them after the equipment is in place. Structural modification, if required, shall be subject to review by the Architect.

## 6. TEMPORARY ELEVATOR

a. Provide a temporary elevator for necessary service during construction operations after the hoistway enclosures are completed and electrical power is available; use temporary machines, or at the Contractor's option, use permanent machines, if they are available in due time for the required services.

b. The temporary elevator shall include temporary wood cars with suitable gates, including temporary hoistway doors, all designed in accordance with the local and state safety requirements.

c. The temporary services shall include qualified operating and maintenance personnel to perform the work in connection with the temporary operations.

d. Upon completion of temporary use, all worn or damaged parts are to be replaced and all equipment placed in first-class condition equal to new.

## 7. TEMPORARY FENCE

a. Provide and maintain a temporary fence to enclose the area at the job site and to guard and close effectively in the designated area. Provide gates at locations where required for access to the enclosed area. Gates shall be of substantial construction, cross-braced, hung on heavy strap hinges, and shall have suitable hasps and padlocks. Submit shop drawings of fence and gates for review of Architect and Owner. Paint the fence with two coats of an approved paint.

b. Remove the fence upon completion of the work or at such time before final completion as directed by the Owner.

8. TEMPORARY FIELD OFFICES

a. Provide and maintain a field office with a telephone at the job site. In addition, provide a temporary office with not less than 200 ft$^2$ of space for the use of the Architect at the construction site. The Architect's office shall be complete with light, heat, air conditioning, toilet facilities, electric water cooler, plan racks, four-drawer metal file with lock, shelves for samples, tables, chairs, and janitor service. When it becomes possible to establish an office in the building, office accommodation of approximately the same size as those in the field offices, including the services above, shall be provided and maintained for the Architect until the issuance of a certificate of substantial completion. Field and temporary offices shall be removed when no longer required. Provide a telephone for the Architect and pay all charges for installation and calls, including long distance calls.

b. Construction shanties, sheds, and temporary facilities provided as required above, or for the contractor's convenience, shall be maintained in good condition and neat appearance, including painting with two coats of approved paint of a color selected by the Architect.

9. TEMPORARY FIRE STANDPIPE SYSTEM

a. Provide a temporary fire standpipe system in all parts of building for use of Fire Department during construction.

b. Permanent risers shall be installed as floor slabs are cast, with capped 2½ in. hose valves on each floor and temporary cap or plug on top. One riser at a time shall be extended up so that remainder are available for use at all times.

c. Install permanent cross connections or provide temporary cross connections.

d. Provide temporary siamese connected to temporary or permanent cross connections.

e. Install one fire water service and one domestic water service at start of project.

f. Obtain early delivery of fire pump and one large domestic booster pump (provide temporary motor controller for booster pump), and install same in permanent or temporary location so that they can be used for temporary fire protection when zone of system they serve requires them.

g. Provide necessary temporary heated enclosures around pumps and necessary heating of service piping to pumps to prevent freezing.

h. System shall be maintained dry during freezing conditions.

i. Provide temporary hose and nozzles as required by Fire Department.

10. TEMPORARY HEAT AND VENTILATION

a. Provide temporary heat as required during construction to protect the work from freezing or frost damage, and as necessary to ensure suitable working conditions for the construction operations of all trades. In areas of the building where work is being conducted, the temperature shall be maintained as specified in the various sections of the specifications, but not less than 45°F. Under no circumstances shall the temperature be allowed to reach a level that will cause damage to any portion of the work which may be subject to damage by low temperatures.

b. Until the building, or any major portion thereof, is enclosed, temporary heating shall be by smokeless portable unit heaters of type listed by Underwriter's Laboratories, Factory Mutual, and the Fire Marshall. Pay for fuel, maintenance, and attendance required in connection with the portable unit heaters. Interior or exterior surfaces damaged by the use of these space heaters shall be replaced by new materials or be refinished.

c. The building shall be considered enclosed when it has reached the stage when exterior walls have been erected, the roof substantially completed, exterior openings closed up either by the permanently glazed windows and doors or by adequate temporary closing, and the building is ready for interior masonry and plastering operations.

d. After the building, or any major portion thereof, has been enclosed the permanent heating system as specified below may be used for temporary heat.

e. When the permanent heating system, or a suitable portion thereof, is in operating condition, the system may be used for temporary heating, provided that the Contractor (1) obtains approval of the Architect; (2) assumes full responsibility for the entire heating system; and (3) pays all costs for fuel, operation, maintenance, and restoration of the system.

f. Provide adequate ventilation as required to keep the temperature of the building within 10°F of the ambient outdoor temperature when such ambient temperature exceeds 70°F, and to prevent accumulation of excess moisture or to prevent excess thermal movement in the building.

g. When the permanent air circulation system, or a suitable portion thereof, is in operating condition, it may be used without refrigeration or chilling, provided that the Contractor (1) obtains approval from the Architect; (2) assumes full responsibility for the system which he is using; and (3) pays costs for power, operation, maintenance, and restoration of the system. Provide temporary filters to adequately filter air being distributed through the duct work to the supply outlets; disposable filters shall be placed in front of all exhaust registers to keep construction dirt out of exhaust duct work. The Contractor shall thoroughly clean the interior of the air handling units and duct work prior to acceptance of the work.

h. Upon conclusion of the temporary heating period, remove all temporary piping, temporary heating units, or other equipment and pay all costs in connection with repairing any damage caused by the installation or removal of temporary heating equipment. Thoroughly clean and recondition those parts of permanent heating and air circulation systems used for temporary service.

11. TEMPORARY LIGHT AND POWER

a. Make all arrangements with the local electric company for temporary electrical service to the construction site; provide all equipment necessary for temporary power and lighting;

and pay all charges for this equipment, the installation thereof, and for current used. The electrical service shall be of adequate capacity for all construction tools and equipment without overloading the temporary facilities and shall be made available for power, lighting, and construction operations of all trades.

b. In addition to the electrical service, provide power distribution as required throughout structure of 120/208 V, three-phase, four-wire, 60 cycle, ac. The terminations of power distribution shall be at convenient locations in the building. Terminations shall be provided for each voltage supply complete with circuit breakers, disconnect switches, and other electrical devices as required to protect the power supply system.

c. A temporary lighting system shall be furnished, installed, and maintained as required to satisfy minimum requirements of safety and security. The temporary lighting system shall afford general illumination in all building areas and shall supply not less than $1W/ft^2$ of floor area for illumination in the areas of the building where work is being performed.

d. All temporary equipment and wiring for power and lighting shall be in accordance with the applicable provisions of the governing codes. All temporary wiring shall be maintained in a safe manner and utilized so as not to constitute a hazard to persons or property.

e. When the permanent electrical power and lighting systems are in operating condition, they may be used for temporary power and lighting for construction purposes, provided that the Contractor (1) obtains the approval of the Architect; (2) assumes full responsibility for the entire power and lighting systems; and (3) pays costs for power, operation, maintenance, and restoration of the systems.

f. At the completion of the construction work all temporary wiring, lighting, and other temporary electrical equipment and devices shall be removed.

12. TEMPORARY ROADS AND ACCESS TO SITE
   a. Construct and maintain in good usable condition all required

temporary roads and access to site, and, when no longer required, remove all temporary construction and restore the site.

b. Access to the site for delivery of construction material or equipment shall be made only from locations designated by the Owner.

13. TEMPORARY STAIRS, LADDERS, RAMPS, RUNWAYS, AND SO ON

    a. Provide and maintain all equipment such as temporary stairs, ladders, ramps, runways, chutes, and so on, as required for the proper execution of the work.

    b. All such apparatus, equipment, and construction shall meet all requirements of the Labor Law and other state or local laws applicable thereto.

    c. As soon as permanent stairs are erected, provide temporary protective treads, handrails, and shaft protection.

14. TEMPORARY TOILETS

    a. Provide and maintain in a sanitary condition enclosed weathertight toilets for the use of all construction personnel at a location within the contract limits, complete with fixtures, water, and sewer connections and all appurtenances. Upon completion of the work, toilets and their appurtenances shall be removed. Installation shall be in accordance with all applicable codes and regulations of authorities having jurisdiction. Chemical toilets will be permitted. The number of toilet rooms required shall be in accordance with the ANSI *Standard Safety Code for Building Construction* or other local authorities.

15. TEMPORARY WATER SERVICE

    a. Provide at a point within 10 ft of the building (or buildings) all water necessary for construction purposes. Make all temporary connections to existing mains; provide temporary meter; and make arrangements and pay for the temporary water service including cost of installation, maintenance thereof, and water used.

    b. Furnish drinking water with suitable containers and cups for use of employees. Drinking water dispensers shall be conveniently located in the building where work is in progress.

c. When the permanent water supply and distribution system has been installed, it may be used as a source of water for construction purposes, provided that the Contractor (1) obtains the approval of the Architect; (2) assumes full responsibility for the entire water distribution system; and (3) pays costs for operation, maintenance, and restoration of the system including the cost of water used.

d. At the completion of the construction work or at such time after the Contractor makes use of the permanent water installation, all temporary water service equipment and piping shall be removed and all worn or damaged parts of the permanent system shall be replaced and equipment placed in first class condition equal to new.

16. SECURITY

a. Provide sufficient watchman service to prevent illegal entry or damage during nights, holidays, or other periods when work is not being executed, and such other control watchmen as required during working hours.

b. Provide all temporary enclosures required for protecting the project from the exterior, for providing passageways, for the protection of openings both exterior and interior, and any other location where temporary enclosures and protection may be required.

c. Take adequate precautions against fire; keep flammable material at an absolute minimum; and ensure that such material is properly handled and stored. Except as otherwise provided herein, do not permit fires to be built or open salamanders to be used in any part of the work.

17. WATER AND SNOW CONTROL

a. From the commencement to the completion of the work, keep all parts of the site and the project free from accumulation of water, and supply, maintain, and operate all necessary pumping and bailing equipment.

b. Remove snow and ice as necessary for the protection and prosecution of the work, and protect the work against weather damage.

SECTION 01600
PRODUCTS

# 1. TRANSPORTATION AND HANDLING

a. Materials, products, and equipment shall be properly containerized, packaged, boxed, and protected to prevent damage during transportation and handling.

b. More detailed requirements for transportation and handling are specified under the technical sections.

# 2. STORAGE AND PROTECTION

a. Provide suitable temporary weathertight storage facilities as may be required for materials that will be damaged by storage in the open.

b. Available storage space at the job site is limited to the site shown on the Drawings. Allocate such space for storage purposes. Any additional off-site space required is the responsibility of the Contractor.

c. Allocate the available storage areas and coordinate their use by the trades on the job. Maintain a current layout of all storage facilities.

d. Store and protect materials delivered at the site from damage. Do not use damaged material on the work.

# 3. INSTALLATION REQUIREMENTS

a. Manufactured articles, materials, and equipment shall be applied, installed, connected, erected, used, cleaned, and conditioned as directed by the respective manufacturers, unless otherwise specified.

# 4. IDENTIFYING MARKINGS

a. Name plates and other identifying markings shall not be affixed on exposed surfaces of manufactured items installed in finished spaces.

# 5. PRODUCT APPROVAL STANDARDS

a. *Definitions*

1. The term product shall include material, equipment, assembly methods, manufacturer, brand, trade name, or other description.

2. References to approved equal or similar terms mean that approval of the Architect is required.

b. *Proof of Compliance.* Whenever the Contract Documents require that a product be in accordance with Federal specification, ASTM designation, ANSI specification, or other association standard, the Contractor shall present an affidavit from the manufacturer certifying that the product complies therewith. Where requested or specified, submit supporting test data to substantiate compliance.

c. *Inclusion in Specifications of Nonspecified Products Prior to Bid Date.* For inclusion of products other than those specified, Bidders shall submit a request in writing at least ten days prior to bid date. Requests received after this time will not be reviewed or considered regardless of cause. Requests shall clearly define and describe the product for which inclusion is requested. Inclusion by the Architect will be in the form of an addendum to the Specifications issued to all contract bidders on record.

d. *Substitutions After Award of Contract*

   1. Substitution of products will be considered after award of contract only under one of the following conditions:

      (a) When the specified product is not available, a proposed substitution will not be considered unless proof is submitted that firm orders were placed within ten (10) days after review by the Architect of the item listed in the Specifications or the unavailability is due to a strike, lockout, bankruptcy, discontinuance of the manufacture of a product, or natural disasters.

      (b) When a guarantee of performance is required and, in the judgment of the Contractor, the specified product or process will not produce the desired results.

   2. Request for such substitution shall be made in writing to the Architect within ten (10) days of the date that the Contractor ascertains he cannot obtain the material or equipment specified, or that the performance cannot be guaranteed.

e. *Procedures Respecting Substitution*

   1. The Contractor shall accompany any request for substitution with such drawings, specifications, samples, manufacturer's

literature, performance data, and other information necessary to describe and evaluate the proposed substitution completely. The burden of proof shall be on the contractor.

2. Permission to make any substitution after award of contract shall be effected by a Change Order. It shall not relieve the Contractor, any subcontractor, manufacturer, fabricator, or supplier from responsibility for any deficiency that may exist in the substituted product or for any departures or deviations from the requirements of the contract documents as modified by such Change Order. Except as otherwise expressly specified by the Contractor in his request for substitution and expressly approved in such Change Order, the Contractor shall be deemed to warrant by his request that the proposed substitute product will satisfy all standards and requirements satisfied by the originally specified product and the Change Order shall not be deemed to modify the Contract Documents with respect thereto.

3. If any substitution will affect a correlated function, adjacent construction, or the work of other trades or contractors, the necessary changes and modifications to the affected work shall be considered as an essential part of the proposed substitution, to be accomplished by the Contractor without additional expense to the Owner if and when accepted.

## SECTION 01700
## PROJECT CLOSEOUT

1. CLEANING UP

  a. The premises and the job site shall be maintained in a reasonably neat and orderly condition and kept free from accumulations of waste materials and rubbish during the entire construction period. Remove crates, cartons, and other flammable waste materials or trash from the work areas at the end of each working day.

  b. Elevator shafts, electrical closets, pipe and duct shafts, chases, furred spaces, and similar spaces that are generally unfinished shall be cleaned and left free from rubbish, loose plaster, mortar drippings, extraneous construction materials, dirt, and dust.

  c. Rubbish shall be lowered by way of chutes, taken down on

hoists, or lowered in receptacles. Under no circumstances shall any rubbish or waste be dropped or thrown from one level to another within or outside the building.

d. Care shall be taken by workmen not to mark, soil, or otherwise deface finished surfaces. In the event that finished surfaces become defaced, clean and restore such surfaces to their original condition.

e. Clean up immediately upon completion of each trade's work.

f. Clean areas of the building in which painting and finishing work is to be performed just prior to the start of this work, and maintain these areas in satisfactory condition for painting and finishing. This cleaning includes the removal of trash and rubbish from these areas, broom cleaning of floors, the removal of any plaster, mortar, dust, and other extraneous materials from finish surfaces, including but not limited to, exposed structural steel, miscellaneous metal, woodwork, plaster, masonry, concrete, mechanical and electrical equipment, piping, ductwork, conduit, and also surfaces visible after permanent fixtures, induction unit covers, convector covers, covers for finned tube radiation, grilles, registers, and other such fixtures or devices are in place.

g. In addition to the cleaning specified above and the more specific cleaning that may be required in various sections of the Specifications, the building shall be prepared for occupancy by a thorough cleaning throughout, including washing (or cleaning by other approved methods) of surfaces on which dirt or dust has collected, and by washing glass on both sides. Wash exterior glass using a window cleaning contractor specializing in such work. Provide and maintain adequate runner strips of nonstaining reinforced Kraft building paper on finished floors as required for protection. Leave equipment in an undamaged, bright, clean, and polished condition. Recleaning will not be required after the work has been inspected and accepted unless later operations of the contractor make recleaning of certain portions necessary.

h. Upon completion of the work, remove temporary buildings and structures, fences, scaffolding, surplus materials, and rubbish of every kind from the site of the work.

## 2. DOCUMENTS REQUIRED PRIOR TO FINAL PAYMENT

a. Prior to final payment, and before the issuance of a final certificate for payment in accordance with the provisions of the General Conditions, file the following papers with the Architect:

1. *Guarantees.* The guarantee required by the General Conditions and any other extended guarantees stated in the technical sections of the Specifications.

2. *Release or Waiver of Liens.* As required by the General Conditions.

3. *Operation and Maintenance Manuals*

   (a) Furnish three (3) complete sets of manuals containing the manufacturers' instructions for maintenance and operation of each item of equipment and apparatus furnished under the Contract and any additional data specifically required under the various sections of the Specifications.

   (b) Arrange the manuals in proper order, indexed and suitably bound. Certify by endorsement thereon that each of the manuals is complete and accurate. Assemble these manuals for all divisions of the work, review them for completeness, and submit them to the Architect. Provide suitable transfer cases and deliver the manuals therein, indexed and marked for each division of the work.

4. *Project Record Documents*

   (a) As the work progresses keep a complete and accurate record of changes or deviations from the Contract Documents and the shop drawings, indicating the work as actually installed. Changes shall be neatly and correctly shown on the respective portion of the affected document, using blackline prints of the Drawings affected, or the Specifications, with appropriate supplementary notes. This record set Drawings, shop drawings, and Specifications shall be kept at the job site for inspection by the Architect and Owner.

   (b) The records above shall be arranged in order, in accordance with the various sections of the specifications, and properly indexed. At the completion of the work, certify by endorsement thereof that each of the revised prints of the Drawings and Specifications is complete and accurate. Prior to application for final payment, and as a condition to its approval by the Architect and Owner, deliver the record Drawings and Specifications, arranged in proper order, indexed, and endorsed as hereinbefore specified. Provide suitable transfer cases and deliver the records therein, indexed and marked for each division of the work.

(c) No review or receipt of such records by the Architect or Owner shall be a waiver of any deviation from the Contract Documents or the shop drawings or in any way relieve the Contractor from his responsibility to perform the work in accordance with the Contract Documents and the shop drawings to the extent they are in accordance with the Contract Documents.

5. *Certificate of Occupancy.* Where the local law at the site of the building requires either a temporary or permanent Certificate of Occupancy, obtain and pay for these cirtificates and deliver to the Architect.

## SECTION 01800
## ALTERATIONS

1. SCOPE
   a. *Work Included.* Perform alterations and related work as shown or specified and in accordance with requirements of the Contract Documents.
   b. *Work of Other Sections*
      1. Reinforcing, cutting, and other modifications of existing structural steel.
      2. Disconnecting, removal, and/or relocation of existing mechanical and electrical work, including equipment, piping, wiring, and so on.
2. STANDARDS
   a. Except as modified by governing codes and by this Specification, comply with the applicable provisions and recommendations of ANSI A10.2, *Safety Code for Building Construction.*
3. SCHEDULING
   a. Before commencing any alteration or demolition work, submit

for review by the Architect and approval of the Owner, a schedule showing the commencement, the order, and the completion dates for the various parts of this work.

b. Before starting any work relating to existing utilities (electrical, sewer, water, heat, gas, fire lines, etc.) that will temporarily discontinue or disrupt service to the existing building, notify the Architect and the Owner 72 hr in advance and obtain the Owner's approval in writing before proceeding with this phase of the work.

4. PROTECTION

a. Make such explorations and probes as are necessary to ascertain any required protective measures before proceeding with demolition and removal. Give particular attention to shoring and bracing requirements so as to prevent any damage to existing construction.

b. Provide, erect, and maintain catch platforms, lights, barriers, weather protection, warning signs, and other items as required for proper protection of the workmen engaged in demolition operations, occupants of the building, public, and adjacent construction.

c. Provide and maintain weather protection at exterior openings so as to fully protect the interior premises against damage from the elements until such openings are closed by new construction.

d. Provide and maintain temporary protection of the existing structure designated to remain where demolition, removal, and new work is being done, connections made, materials handled, or equipment moved.

e. Take necessary precautions to prevent dust and dirt from rising by wetting demolished masonry, concrete, plaster, and similar debris. Protect unaltered portions of the existing building affected by the operations under this section by dustproof partitions and other adequate means.

f. Provide adequate fire protection in accordance with local Fire Department requirements.

g. Do not close or obstruct walkways, passageways, or stairways without the authorization of the Architect. Do not store or place materials in passageways, stairs, or other means of egress. Conduct operations with minimum traffic interference.

h. Be responsible for any damage to the existing structure or contents of the insufficiency of protection provided.

5. WORKMANSHIP

   a. Demolition, removal, and alteration work shall be as shown on the drawings. Perform such work required with due care, including shoring, bracing, and so on. Be responsible for any damage that may be caused by such work to any part or parts of existing structures or items designated for reuse. Perform patching, restoration, and new work in accordance with applicable technical sections of the Specifications.

   b. Materials or items designated to become the property of the Owner shall be as shown on the drawings. Remove such items with care and store them in a location at the site to be designated by the Owner.

   c. Materials or items designated to be reinstalled shall be as shown on the Drawings. Remove such items with care under the supervision of the trade responsible for reinstallation; protect and store until required. Replace material or items damaged in its removal with similar new material.

   d. Materials or items demolished and not designated to become the property of the Owner or to be reinstalled shall become the property of the Contractor and shall be removed from the Owner's property.

   e. Execute the work in a careful and orderly manner, with the least possible disturbance to the public and to the occupants of the building.

   f. In general, demolish masonry in small sections. Where necessary to prevent collapse of any construction, install temporary shores, struts, or bracing.

   g. Where alterations occur, or new and old work join, cut, remove, patch, repair, or refinish the adjacent surfaces or so much thereof as is required by the involved conditions, and leave in as good a condition as existed prior to the commencing of the work. The materials and workmanship employed in the alterations, unless otherwise shown or specified, shall conform to that of the original work. Alteration work shall be performed by the various respective trades that normally perform the particular items of work.

h. Finish new and adjacent existing surfaces as specified for new work. Clean existing surfaces of dirt, grease, loose paint, and so on, before refinishing.

i. Where existing equipment and/or fixtures are indicated to be reused, repair such equipment and/or fixtures and refinish to put in perfect working order. Refinish as directed.

j. Cut out embedded anchorage and attachment items as required to properly provide for patching and repair of the respective finishes.

k. Confine cutting of existing roof areas designated to remain to the limits required for the proper installation of the new work. Cut and fold back existing built-up roofing. Cut and remove insulation, and so on. Provide temporary weathertight protection as required until new roofing and flashings are installed. Consult the Owner to ascertain if existing guarantee bonds are in force and execute the work so as not to invalidate such bonds.

6. CLEANING UP

a. Remove debris as the work progresses. Maintain the premises in a neat and clean condition.

# 13

## Specifying Materials

The selection of materials and equipment in the design of a structure is the responsibility of the architect. His professional judgment dictates the quality of the item to be specified. He is similarly responsible for selecting materials for use in conjuction with other materials or assemblies of materials and equipment in a composite design. Inasmuch as he is held accountable for the success or failure of his plans and specifications, he should as a logical consequence be the master of his own fate and have ultimate control in this selection.

In many instances, standards for materials have been established by certain recognized authorities. They include ASTM Standards, Federal Specifications, ANSI Standards, AASHO Specifications, and Product Standards. These standards establish various types, grades, and qualities, and, in addition, may offer many options. The standards can be used if found satisfactory by the specifier, or he can upgrade their requirements by specifying additional characteristics. It is also customary and quite necessary to use trade or brand names in specifying materials when reference standards have not been developed and when, in the judgment of the architect, these brands or proprietary materials will fulfill the project requirements.

When brand names are used as a standard in a specification, it is almost impossible to include the names of all competitive materials that the architect may be willing to use. Competition is invited in order to obtain equitable costs to the owner. To allow for the possible use of other brands or makes without naming them in endless profusion, it has been the custom to follow the name given in the specifications with the words "or equal." This device has often led to conflict between architect and contractor concerning who should determine the equality of materials proposed for substitution. Undoubtedly, no phrase in specifications has been subject to more severe criticism than the phrase "or equal." That the use of this term is not satisfactory in controlling the selection of materials and equipment specified is attested to by the problems that have arisen from its use, by the countless seminars that have been held to discuss alternative approaches, and by the many articles that have appeared over the years in attempts to arrive at a more satisfactory solution.

As a result, several other systems are in use today. Descriptions of them are set forth on the following pages, beginning with a summary of the disadvantages of the traditional or equal specifications.

## *Or Equal Specifications*

Or equal specifications usually name one, two, or several brand names and follow with the term "or equal" or "or approved equal." The following are some of the reasons that have been advanced for eliminating the term "or equal" from specifications:

1. When the or equal phrase is used, a bidder attempts to secure a lower price on a material than that specified, and he will be in doubt as to whether the architect will approve it. If the bidder takes a chance on this lower price material, he risks being forced to buy the higher priced material specified. If the bidder does not take this chance, he loses the advantages of the lower price, which might make the difference between winning or losing the contract.

2. The or equal clause increases the amount of office work the architect must perform in order to investigate all the or equal substitutions that are submitted by the contractor for approval.

3. It permits the contractor more opportunity for last-minute substitutions, requiring overhasty consideration by the architect.

4. Where a continuing project is developed intwo or more phases of construction, the or equal clause may allow different materials to be used in the same project, and the maintenance problems of the owner are multiplied.

5. Although an alternate product may be equal or similar to the one specified, its use in conjunction with other assemblies, materials, or products may be unproven, unacceptable, improper, or even faulty. At times, additional costs are incurred while making adaptations to accommodate the alternate product, and such costs are difficult to resolve. While the architect derives no benefit in making adaptations, it often has caused additional expense to the architect in connection with the changes.

6. It contributes heavily to "bid shopping," which results in delays in construction, since the substitution is usually submitted at the last moment and interferes with the routine process of careful evaluation by the architect.

7. It takes control of the project away from the architect who is responsible for its execution.

### Open Specifications

The open specification for materials and equipment is written without reference to brand names or proprietary marks. This type of specification is used most commonly for public work, and can be used in private work. It can be written quite simply by the use of reference specifications which make reference to recognized standards discussed in Chapter 6. In the absence of standards for some items, notably for mechanical and electrical equipment, a descriptive or performance specification must be prepared to present complete and comprehensive data on the product required. The open specification is intended to invite the greatest amount of competition and to maintain complete impartiality between various manufacturers.

The open specification can be used for many basic materials when reference is made to recognized standards. These standards include such materials as structural steel, cement, gypsum, ceramic tile, concrete masonry units, roofing felts and bitumen, and a host of other materials. It becomes inadequate when the architect seeks to specify paints, many sealants, concrete admixtures, elastomeric waterproofing materials, and, generally, the man-made products of chemistry, since the promulgation of adequate standards lags far behind the development of these products. Standards are similarly inadequate for specifying such equipment as boilers, lighting fixtures, fans, pumps, and other items of a mechanical or electrical nature. For these products, the architect and engineer contrive to write an open specification setting forth descriptive or performance characteristics that create voluminous specifications.

When reference specifications are used, the architect and engineer can approve submissions quite readily by requiring certifications from manufacturers attesting to compliance of their products with the standards specified. However, when descriptive or performance specifications for materials and equipment are employed, the architect and engineer must carefully check the submission against all the provisions that have been minutely specified. Failure to check against compliance with even one aspect of his own specifications, which may result in a subsequent failure, can lead to legal action against him by the owner for having approved a product which did not comply with the specifications.

## Base Bid or Closed Specifications

A base bid or closed specification is one in which the architect specifies only one brand name or proprietary make for each individual material, piece of equipment, or product. Occasionally he may augment this brand name with a brief descriptive specification or cite performance characteristics. The intent of this type of specification is to limit the bidding to products that the architect or engineer has specifically selected for the project. The bidder has no choice under this base bid specification.

Under this system, product selection and responsibility rest entirely with the architect. It enables the architect to set room sizes, headroom, and vital dimensions, clearances, and foundations, especially for mechanical equipment. In addition, bid shopping is eliminated, which does away with unnecessary construction delays that are a by-product of this practice.

Under this system, competition is excluded, and the owner does not necessarily get the best value for his dollar. The architect is sometimes unfairly accused of favoritism by the manufacturer or supplier not included in the specifications. The contractor is compelled to use the product of a manufacturer or a supplier with whom he does not regularly do business and may experience difficulty with credit and delivery. In addition, the contractor may not have had experience in the installation of the specific product named and may be required to guarantee an installation with which he has had no previous experience.

## Bidders Choice or Restricted Specifications

The bidders choice or restricted specification is akin to the base bid or closed specification, except that the architect names two or more brand names or proprietary makes for each item he wishes to use.

The architect should investigate each of the products he proposes to specify, to make certain that like or equal products are put into competition with one another. It has an advantage over the base bid specification in that competition is invited. The architect must be careful not to equate several materials where one is so much lower in price that it negates the advantage of competitive bidding and, in effect, creates a closed specification.

## Bidders List of Substitutions

Under this method, the bidder is permitted to submit alternatives or substitutions for the materials or equipment specified. These substitutions are listed and included with his bid, along with the net difference in cost if the substitution is accepted. Generally, the bid must also include the name, brand, catalog number, and manufacturer of the proposed substitute, together with complete specifications and descriptive data.

When one calculates the hundreds of items used in the construction of a building and the quantity of products manufactured, the number of substitutions can be staggering. An evaluation and analysis of bids to determine an acceptable low bidder would be a Herculean task, and the bids would no longer be predicated on the architect's original selection of materials and equipment.

This method does not entirely achieve the element of competition. Since each bidder is free to submit any substitution, and since each of the bidders is unaware of what substitutions his competitor may offer, there is no competition on the substitutions offered.

### Product Approval Standards

Under this method, products are clearly defined by using specification standards where possible, by using specific product names, by specifying more than one product where possible, and by listing basic criteria where desirable. Bids are based on the use of any product meeting established standards (such as ASTM) or the products specified. However, upon application, bidders are permitted to request approval of products during the bidding period, within established limits. If the architect approves the product, it is listed in an addendum so that all bidders compete on the same basis.

It is recognized that some of the disadvantages of methods previously discussed can be applied to this method. However, this method does achieve the following:

1. Control of products by the architect.
2. Competition.
3. Fairness in attracting other products of which the architect may not have been aware at the time of preparation of documents.
4. Elimination of the risk to bidders in accepting products other than those specified.
5. Closer bidding.
6. Discouragement of bid peddling or shopping.
7. Administration of equality at the proper time and by the proper agency.
8. Complete flexibility.

Under the Product Approval Standards, manufacturers of materials and equipment receive consideration under competitive bidding procedures. The auction is over when the bids are submitted. The possibility of the successful contractor submitting after award of a contract (under the or equal method) material or equipment not previously known to all bidders, and which in effect prevents the owner from obtaining competitive prices, is precluded.

This method lends itself to use in projects involving public funds. According to the U.S. General Accounting Office, this procedure can be used by federal agencies if they so decide. Legal officers in state, county, and municipal governments can similarly be apprised of this solution and may give approval to this method so that the use of the term "or equal," as it has been used and abused, is placed in proper perspective.

It should be noted that the intent of equality to obtain competition is not changed or excluded from the specification, but rather the time for evaluation is adjusted so as to occur prior to the bid date.

The greatest apparent drawback is the possibility of a substantial number of requests for approval, within a limited time. Carefully prepared specifications, a slightly extended bidding period, and allotment of sufficient time prior to bid date as the deadline for approvals minimizes the problem.

Control of the project is achieved so that at the time bids are received there is no doubt about the quality of products to be used. Competition is obtained through the basic qualification. In addition, other products that prove to be acceptable can further increase competition. No one with an acceptable product need be kept out. Prime contract bidders can be confident of the product bids they are using in their estimates. Less gamble, less contingency, and sharper bidding are the results.

Bid peddling and shopping are reduced, resulting in more competitive bids from product suppliers. The competitive products are known, and it is also known that a lower priced nonspecified product cannot be offered, and possibly accepted, after bids are received. Therefore, the best price possible consistent with the desired quality can be offered as a firm bid, with reasonable assurance that the price will not be undercut by an unknown or "nonequal." This method removes the problem of bidders claiming that a certain product is equal, and that it was used in their bid. The contractor is not damaged financially as a result of a rejected product. Complete flexibility can be easily realized. Recognized standards can be used completely if available. One product or twenty can be specified, depending on criteria and the owner's requirements, and every element of design and function can be considered. Conditions can be varied from project to project, and from public work to private work. The length of time for approvals can be varied.

The same care is essential with this method as it is with most others. Products must be properly and clearly specified. The basis for evaluation of products must be stated. Proposed substitutions must be given complete consideration, careful review, and honest evaluation.

### Administration of Product Approval Standards

State or specify the conditions only once under Section 01600, Products (see Chapter 12); this establishes the conditions of consideration. The term "or equal" is omitted under individual specification sections or detailed requirements. This forces a bidder to refer to the proper article regarding approval. List all known products acceptable for the project. This is not an overwhelming task; a file can soon be built up to reduce the bulk of the work for most items. Wherever possible use only established standards—such as ASTM, Federal Specifications, and so on—that have been determined acceptable for the project, modifying them where necessary. Insofar as possible, list basic criteria that must be met for product consideration. However, meeting the criteria may not always qualify a product because of intangibles and variables.

Written requests are essential and should be mandatory for the following reasons:

1. They form a basis of understanding in the event of a later claim of misinterpretation.

2. Endless worthless hours on the telephone with persons who are reasonably sure that their product does not comply, but who feel that a phone try is worthwhile, are eliminated.

3. Written requests are generally submitted only by persons with a genuine interest in bidding.

4. They permit a review and evaluation in the quiet of normal office procedure, without the pressure of a sales pitch.

5. For a given level of quality, they weed out requests for products that are obviously below requirements. After an unsuccessful attempt or two, the person making the request stops trying.

Requests should be considered only from prime bidders. Time must be allotted for review and evaluation of requests. This may be difficult at first. However, time always has to be made available for such a review after bids are received, and the process is merely reversed to prebid time. This method must be administered and enforced with a strong will. Deviations cannot be permitted (even to friends in the industry). If the product is not specified, it cannot be used.

Occasionally an equal product is omitted, possibly by oversight. This does not change the conditions, and if the manufacturer does not find his product in the specification, he should request that it be included by one of the prime bidders. Each job and each owner's prejudices are individual considerations. A manufacturer cannot assume that he is approved.

In fairness to bidders, prompt consideration should be given. Addenda should be issued as the bidding period progresses so that those who make early application can know whether they are approved or not, in time for their "take off."

## Specifying Product Approval Standards

The language to be incorporated in Section 01600 if product approval standards are to be used for specifying materials is illustrated in Chapter 12, General Requirements.

# 14

## Specification Language,

It is not intended, nor indeed is it possible, for this chapter to be a treatise on English grammar and readable writing. Rather, it is proposed that we direct ourselves to the reasons why it is necessary to use proper specification language. Each statement in a specification carries a dollar sign alongside it, whether it is concerned with specifying materials, instructing a contractor on installation procedures, or describing workmanship. The contractor expects to be paid for each order given him by the specifier, and the contractor's bid reflects every statement in the specifications. Using vague, ambiguous language indicates that the specifier may want something but is unsure about demanding it. Statements such as "tests will be required unless waived," "additional shop drawings and samples may be required," and "uneven surfaces may be cause for rejection" are examples of equivocation that plague the contractor. Specification language should be precise, not vague. The precise specification can be enforced; the vague one may be difficult to enforce and will still cost the owner money because the contractor has included the cost in his bid.

The essential requirement for writing specifications, aside from technical know-how, is the ability to express one's self in good English. Although the specifications are one of the contract documents that becomes a legal document, legal phraseology is not necessary. A statement in good, clear English may be even more definite, unequivocal, and understandable to the superintendent and the foreman than legal wording.

Language is a means of communication. Unlike graphic communication, where symbols and crosshatching have precise meanings, words must be carefully selected to transmit information. There are subtle variations in the choice of language, and the word or term selected to communicate an instruction may be interpreted by a contractor quite differently from that which was intended by the specifier.

Consider the word "smooth." The dictionary defines it as "having an even surface; devoid of surface roughness." The term "smooth" has been employed in specifications as follows: "Bituminous road surfaces shall be smooth," yet a preferred texture for the surface to reduce skidding is a rough texture. Concrete floors have been specified to have a smooth, wood float finish; or a smooth, rubbed finish; or a smooth, trowelled finish. However, in each case the degree of smoothness varies. It would be preferable to select the tool that will accomplish a result and rely on it to achieve the surface finish desired by specifying its use as follows: "Finish concrete floors with a wood trowel," or "rub concrete steps with carborundum stone," or "steel trowel concrete floors."

In order to communicate by language, the architect should visualize grammar as well as he perceives design. Grammar is not just obeying rules; it is the power politics of language. Words rule other words; subjects have objects. Prepositions are powerful indicators, instruments of authority, and traffic directors. All this suggests the visual, and grammar should be visualized as much as possible.

There are three important Cs for the specification writing. The wording of specifications should be clear, correct, and concise: clear so there is no ambiguity; correct technically; and concise so there is no excessive verbiage. A good specification is one containing the fewest words that can be used to complete the description and make sense. Verbosity and repetition lead to ambiguity.

Considering those who are to use the specifications, it is evident that specifications must be made clear to some whose vocabulary may be limited. The meaning should be grasped readily even by the workmen. A specification written in English that is clear even to the mechanics on the job is the logical form to use. If a mechanic cannot interpret specifications, he will not be able to execute them.

Since specifications are instructions to the contractor, they should be definite and mandatory. To be mandatory they must be imperative. Therefore use the imperative "shall" with reference to the work of the contractor and never use the vague and indefinite "will" or "to be." The proper place to use "will" is in a statement describing the acts of the owner or the architect.

Some examples of specification language are contained in a series of *Maxims for Specification Writing* by the late H. Griffith Edwards, FAIA, FCSI, as follows:

### *Maxim No. 2   Be Brief*

Specifications tend to be too lengthy in spite of the greatest economy of words. A constant effort should be made by all specification writers to say the same thing with less verbiage and in a condensed manner. Avoid long and involved sentences.

The following submaxims will help in this regard:

a. *Specify Standard Articles by Reference to Accepted Standard Specifications*: For example: Many words would be necessary to describe properly a common product such as portland cement; its chemical composition, fineness, soundness, compressive strength, tensile strength, and so on, should be mentioned. All these words are eliminated by a simple reference to the standard, thus:
   *Portland Cement* shall meet the requirements of ASTM Spec. C 150, Type I.

b. *Avoid Repetition of Information* shown or scheduled on the drawings. Also, avoid duplication within the specifications themselves. This will eliminate words and the possibility of contradiction.

c. *Do Not Include Inapplicable Text*: Avoid discussion of materials or methods that could not pertain to the actual construction work for which a set of specifications is prepared, as it is confusing to bidders. When old project specifications are used for the preparation of the new project specifications, the writer sometimes carelessly overlooks deleting inapplicable material.

d. *Never Make the Word "Contractor" the Subject of a Sentence in a Trade Section.* Instead, make the material or method the subject. This is a maxim from Arthur W. Farrell, Head Specification Writer, 6th Naval District of the Bureau of Yards and Docks, USA. Not only will this make your sentence shorter, but also it will put up front the key word to serve as a title, thus:

Poor:  *Rubbed Finish* Contractor shall apply a rubbed finish to exposed surfaces of concrete.

Better:  *Rubbed Finish* shall be applied to exposed surfaces of concrete.

e. *Eliminate Superfluous Words* such as the following:

*All*:

The use of the word "all" is frequently unnecessary.

Poor:  Store all millwork under shelter.

Better:  Store millwork under shelter.

*Which*:

"Which" and other relative pronouns such as "who" and "that" should be used sparingly, if at all.

Poor:  Install bathroom accessories which are to be purchased under an allowance. . . .

Better:  Install bathroom accessories to be purchased under an allowance. . . .

*The*:

Definite article "the" and indefinite articles "a" and "an" need not be used in many instances. The following paragraph actually reads better with the underscored words deleted:

The Contractor shall strip the top soil from the areas to be excavated and graded, and neatly pile it on the property; then, after all the backfilling is finished and all the areas graded, the available top soil shall be spread over the areas to be seeded or planted.

*Of*:

The preposition "of" may often be eliminated to shorten the text:

Poor:  For colors see Schedule of Paint Finishes.

Better:  For colors see Paint Finish Schedule.

> Poor:    Apply one coat of stipple finish to walls in the Office of the Manager.
>
> Better:  Apply one stipple finish coat to walls in Manager's Office.

f. *Use Numerals Instead of Writing Out Numbers*: The practice of using numerals, rather than writing out the numbers throughout specifications, is recommended for the reason that numerals are used on the drawings and they make for clearer, easier reading, and it shortens the text. Numerals on the drawings, which are part of the contract documents, are considered legally binding and numerals in specifications are similarly legally binding.

> Poor:    Four feet, six inches, Twenty-six gauge. . . .
>
> Better:  4 ft 6 in., 26 gauge. . . .

g. *Use Well-Known and Accepted Abbreviations*: The use of abbreviations facilitates reading, reduces the typing, and shortens the text without sacrificing clarity. The following abbreviations may be used with impunity:

> *General Abbreviations*, such as ASTM, bbl., Co., Corp., cu., Fed. Spec., ft., gals., hr., in., Inc., lb., lin., max., min., o.c., oz., sec., sq., wt.
>
> *Engineering Abbreviations*, such as ACI, AISC, I.D., psi, psf, rd.
>
> *Lumber Abbreviations*, such as AD, Btr., Com., Dim., EM, J&P, KD, M, Mbm, SIS, S2S, S4S, T&G, V1S, V2S, VG.
>
> *Electrical Abbreviations*, such as amp., kw., hp., AC, DC, NEC, AWG.
>
> *Heating Abbreviations*, such as Btu, Btuh, C., F., cfm, fpm, gpm, rpm, EDR, ASHR&ACE.

h. *Use Simple Imperative Mood and Simple Present Infinitive as Often as Possible* to shorten the text. Interpreting specifications as being written instructions addressed to the contractor, which they are, the simple imperative mood is quite appropriate.

> Poor:    Contractor shall install lighting fixtures which will be furnished by Owner.
>
> Better:  Install fixtures to be furnished by Owner.

i. *Consider the Use of Streamlined Specifications*: In *Pencil Points Magazine* in August 1939 there appeared an article entitled

"Streamlined Specifications" by Horace W. Peaslee, FAIA, proposing writing specifications in an outline form without the use of complete sentences. For example, a paragraph in a masonry section would be written as follows:

*MORTAR MATERIALS*:
   *Portland Cement*: ASTM Spec. C 150, Type I.
   *Masonry Cement*: ASTM Spec. C 91, Type II.
   *Slag Cement*: ASTM Spec. C 358.
   *Hydrated Lime*: ASTM Spec. C207. Putty by adding water.
              Store 24 hr before use.
   *Sand*: ASTM Spec. C 144.

### Maxim No. 3   *Use Simple and Clear Language*

. . . which is readily understood by the average layman. Be specific. Avoid the use of indefinite words or clauses. Attempt to prepare specifications that will require no interpretation as to meaning.

Under this maxim there are the following submaxims:

a. *Use "Shall" in Connection With Acts of the Contractor, or With Labor, Materials, or Equipment to be Furnished by Him,* but the use of the simple imperative mood is even better.
   Poor:    Brick will be laid in running bond.
   Better:   Brick shall be laid in running bond.
   Best:    Lay brick in running bond.

b. *Avoid the Use of "Must" and "Is To" and Substitute the Word "Shall" or the Simple Imperative Mood,* to prevent the inference of different degrees of obligation.
   Poor:    Each joint must be filled solid with mortar.
   Poor:    Each joint is to be filled solid with mortar.
   Better:   Each joint shall be filled solid with mortar.
   Best:    Fill each joint solid with mortar.

c. *Do Not Use "Any" When a Choice Is Not Intended*: For the reason that "any" implies a choice, it should not be used when a choice is not intended, as for example:
   Poor:    Any materials rejected shall be removed.
   Better:   Materials rejected shall be removed.
   Best:    Remove rejected materials.

d. *Do Not Use "Either" When a Choice Is Not Intended*: The

word "both" should be substituted for "either" when no choice is intended.

Poor:    Glass panels shall be installed on either side of main entrance.

Better:    Glass panels shall be installed on both sides of main entrance.

Best:    Install glass panels on both sides of main entrance.

e. *Do Not Use "Same" as a Pronoun*:

Poor:    If materials are rejected, the Contractor shall replace same at no additional cost.

Better:    Replace rejected materials at no additional cost.

f. *Do Not Use "Said" as an Adjective*:

Poor:    Said materials shall be replaced at no additional cost.

Better:    Replace rejected materials at no additional cost.

g. *Do Not Use "and/or"*: This is a stilted legal expression. The word "or" or "both" should be used in place of "and/or."

Poor:    Brick shall be made of clay and/or shale.

Better:    Brick shall be made of clay, shale, or a combination of both.

h. *Do Not Use "Etc."* Placed at the end of a list of items, "etc." shows that the specification writer obviously does not know of what the complete list consists, or he is too lazy to write it out. The use of "etc." is vague, puts unnecessary responsibility on the contractor, and therefore should not be used. As one specification writer puts it, "It is better to be definite even if you are wrong; then, at least, there is a firm basis for negotiating the corrections."

Poor:    All standing trim, running trim, etc., shall be painted.

Better:    Paint exposed millwork.

i. *Do Not Use Phrase "Contractor Shall Furnish and Install"*: Since it is established by the general conditions that the contractor shall provide and pay for all materials, labor, water, tools, equipment, light, power, transportation, and other facilities, unless otherwise stipulated, for the execution and completion of the work, it is redundant to use the phrase in other sections.

Poor:    Contractor shall furnish and install standard size face brick.

Better:    Face brick shall be standard size.

j.  *Do Not Use Phrase "To the Satisfaction of the Architect"* and similar phrases such as "as the Architect may direct," "acceptable to the Architect," and "in the opinion of the Architect." Instead, specify exactly what the architect's directions are, or definitely what would be satisfactory or acceptable to him. Do not leave contractor guessing and at the mercy of architect's future decisions.

Poor:    Brick shall be laid to the satisfaction of the Architect.

Better:    Brick shall be laid plumb and true with all joints completely filled with mortar.

k.  *Do Not Use Phrase "A Workmanlike Job"* and similar phrases such as "a high-class job" and "a first-class job." Instead, the type of workmanship expected should be described in detail.

Poor:    Brick shall be laid in a workmanlike manner.

Better:    Brick shall be laid plumb and true with all joints completely filled with mortar.

## *Maxim No. 5    Make Specifications a Reference Text*

A set of specifications is a reference text and the preparation of an alphabetical cross-index is too involved and complicated to be practical. Furthermore, work should be separated into trade sections reflecting the methods by which work is sublet in the region of the job. Therefore, a logical arrangement of the data covered by the specifications becomes mandatory to facilitate reference and the subletting of work.

a.  *Provide Titles for All Articles*: There is no need to use titles for paragraphs in novels and similar literature, but titles should be provided for articles and paragraphs in reference texts. Since specifications are used extensively for reference, titles should be provided, not only for the articles, but also for the paragraphs. Most specification writers accomplish this by choosing key word or words reflecting the contents, and identifying them as illustrated below:

IA-10    *ARTICLES*: The titles of articles are often capitalized and underlined.

a. *Paragraphs*: Titles of paragraphs are often lowercase letters and underlined.
  1. *Items* may also be lowercase and underlined.
  2. *Numbering Items*: Items are usually numbered 1, 2, 3, and so on.

b. *Capitalize for Easy Reference*: The general rules regarding the capitalization of the first letter of certain words should be followed, but in addition, certain words are written the same as proper names in specifications. They are

  1. *Parties to the Contract*, including Owner and Contractor and those defined in the General Conditions of the Contract including Architect and Subcontractor.
  2. *Spaces of the Building*, such as Principal's Office, Auditorium, Library, Teachers' Lounge, and Clinic.
  3. *The Contract Documents*, including Agreement, General Conditions of the Contract, Supplementary General Conditions, Drawings, and Specifications.
  4. *Grades of Materials*, such as B and Btr southern pine, Intermediate Heat Duty fire clay brick, Standard Grade ceramic tile, and Type I Regular Core hardwood plywood.

c. *Minimize Cross-References* in the specifications to drawings, to specification sections, and to specification articles and paragraphs. When absolutely necessary, do so by referring to titles instead of numbers (numbers are changed more often than titles during developmental stages).

  Poor:    *PILE CAPS* as detailed on Drawing No. S-1 are specified under Section No. 3A.

  Better:  *PILE CAPS* as detailed on Foundation drawing are specified under CONCRETE section.

d. *Do Not Use Long Block Articles*, that is, long unbroken articles covering several phases of one subject. Instead, break the article into paragraphs and give titles to paragraphs for ready reference and better comprehension, as illustrated by the following example:

  Poor:    *TESTS*: Materials used in this work shall be tested by the manufacturer before shipment. Drainage and vent piping shall be tested before fixtures are installed by capping or plugging the openings, filling the entire system with water, and allowing it to stand thus filled for 3 hr. Water supply piping and hot water tanks and heaters inside the building shall be tested by capping or plugging the openings, connecting up a test pump, filling the system with water and applying a hydrostatic pressure of 150 lb/in.². Water piping may be tested before fixtures or faucets are connected. Each fixture shall be tested for soundness, stability of support, and satisfactory operation of all its parts. After fixtures have been installed, all traps shall be filled and a smoke test shall be applied to show up any leaks in the fixtures or connections. Piping shall be absolutely tight under test. Screwed and soldered piping not tight under test shall be taken down and reassembled. Joints in cast iron piping not tight under test shall be replaced with new heaters and tanks. Certificates of tests and final acceptance, to be issued by the local Plumbing Inspector, shall be delivered to the Architect.

Better: *TESTS*: Materials used in this work shall be tested by the manufacturer before shipment.

   a. *Drainage and Vent Piping* shall be tested before fixtures are installed by capping or plugging the openings, filling the entire system with water, and allowing it to stand thus filled for 3 hr.

   b. *Water Supply* piping and the hot water tanks and heaters inside the building shall be tested by capping or plugging the openings, connecting up a test pump, filling the system with water, and applying a hydrostatic pressure of 150 lb/in.². Water piping may be tested before fixtures or faucets are connected.

   c. *Fixtures*: Each fixture shall be tested for soundness, stability of support, and satisfactory operation of all its parts. After fixtures have been installed, all traps shall be filled and a smoke test shall be applied to show up any leaks in the fixtures or connections.

d. *Piping* shall be absolutely tight under test.

e. *Screwed and Soldered Piping* not tight under test shall be taken down and reassembled.

f. *Joints in Cast Iron Piping* not tight under test shall be dug out and joints recaulked and repoured.

g. *Tanks and Heaters* not tight under test shall be replaced with new heaters and tanks.

h. *Certificates of Tests* and final acceptance, to be issued by the local Plumbing Inspector, shall be delivered to the Architect.

# 15

## Specification Reference Sources

The scope of architectural specifications is so broad—encompassing the gamut of materials from acoustical products through zinc coatings—that no single specifier can possibly have a complete and intimate knowledge of all these materials, nor of the constant improvements that manufacturers are continually making. Knowing where to look for information is half the battle; applying that information successfully (see Chapter 16) is the other half. The specifier should familiarize himself with these specification reference sources and learn to differentiate between the good and the not too useful material available, since the latter will simply clutter his files. These technical references should be accumulated and made a part of a ready reference library and file for the specifier's use.

Specification reference sources fall into several categories as follows:

1. Textbooks
2. Materials standards
3. Guide specifications
4. Journals and periodicals
5. Building codes and ordinances
6. Materials investigations
7. Association standards
8. Manufacturers' catalog files
9. General references

## Textbooks

Textbooks serve as a valuable source of information on specification principles. Since specification writing is an art, not a science, the various authors have expressed many opinions that are frequently at variance with one another. The student and the professional specifier can assess for himself the arguments for and against various systems outlined therein, and determine for himself the methods that he will employ in writing his own specifications.

The following textbooks are suggested as a reference source:

H. G. Edwards, *Specifications,* D. Van Nostrand Co.

D. W. Gale, *Specifying Building Construction*, Reinhold Publishing Corporation

G. Goldsmith, *Architects' Specifications and How To Write Them,* American Institute of Architects

H. R. Sleeper, *Architectural Specifications*, John Wiley & Sons, New York

D. A. Watson, *Specifications Writing for Architects and Engineers*, McGraw-Hill Book Co., New York

## Materials Standards

Standards for materials have been devised and issued by governmental agencies, national technical associations, and certain producers of materials in order to provide uniform standards as to criteria, grading, and testing. The specifier should obtain indices listing the various standards promulgated by these agencies and associations in order to secure those which will be useful to him in his work. The specification reference sources for these materials standards are as follows:

*American Society for Testing and Materials*
1916 Race Street
Philadelphia, Pa. 19103
*Federal Specifications*
U.S. Superintendent of Documents
Washington, D.C. 20402

*Simplified Practice Recommendations*
Department of Commerce
U.S. Superintendent of Documents
Washington, D.C. 20402
*Commercial Standards*
Department of Commerce
U.S. Superintendent of Documents
Washington, D.C. 20402
*American National Standards*
American National Standards Institute
1430 Broadway
New York, N.Y. 10018

## Guide Specifications

Guide, or master specifications for technical sections, and technical reports that can serve as an aid in the preparation and development of specifications, can be obtained from the following sources:

*Production Systems for Architects and Engineers*
American Institute of Architects
1785 Massachusetts Avenue, N.W.
Washington, D.C. 20036
*CSI Manual of Practice and Specification Series*
Construction Specifications Institute, Inc.
1150 Seventeenth Street, N.W.
Washington, D.C. 20036

## Journals and Periodicals

Valuable information and discussions on current technical problems written by individuals proficient in various areas of construction materials and techniques are frequently published in the following journals and periodicals:

*Building Research*
Building Research Institute
1725 De Sales Street, N.W.
Washington, D.C. 20036

*Construction Specifier*
Construction Specifications Institute, Inc.
1150 Seventeenth Street, N.W.
Washington, D.C. 20036
*Canadian Building Digest*
National Research Council
Ottawa, Canada

## Building Codes and Ordinances

There are codes and ordinances promulgated by official bodies, cities, and municipalities that have been developed to safeguard health, life, and property. These include zoning regulations, building codes, fire, safety, plumbing, and electrical codes. These codes should be consulted to ensure compliance with them, and by incorporating them into the specifications by reference where necessary. Occasionally these codes may cover only minimum standards and types of construction, and the architect or specifier may prefer to specify a higher grade of construction than that required under the codes. Reference sources often used when no code prevails, or when a better type of construction is desired, are the following codes:

*National Building Code*
American Insurance Association
85 John Street
New York, N.Y. 10038
*Uniform Building Code*
International Conference of Building Officials
50 South Los Robles
Pasadena, Ca. 91101
*Basic Building Code*
Building Officials and Code Administrators International, Inc.
1313 East 60th Street
Chicago, Ill. 60637
*Southern Standard Building Code*
Southern Building Code Congress
1116 Brown-Marx Building
Birmingham, Ala. 35203

*National Electric Code*
National Fire Protection Association
60 Batterymatch Street
Boston, Mass. 02110

## Materials Investigations

Another valuable service that has been performed by some governmental agencies and national technical associations is the laboratory investigation of properties of building materials and the structural elements of buildings, as well as the performance of mechanical equipment for buildings. Many of these reports have also been compiled on the basis of the experience record of many individuals who have had close association with certain materials. These specification reference sources are the following:

> *Building Science Series*
> National Bureau of Standards
> U.S. Superintendent of Documents
> Washington, D.C. 20402
> *Housing Research Papers*
> Housing and Home Finance Agency
> Washington, D.C. 20402
> *The Wood Handbook*
> Forest Products Laboratory
> U.S. Department of Agriculture
> Madison, Wis.
> *Reports*
> Small Homes Council
> University of Illinois
> Urbana, Ill.

## Association Standards

Other very pertinent sources of architectural information can be found in standards issued by various manufacturing, contracting, and technical associations as follows:

*ACI Standards*
American Concrete Institute
Box 4754, Redford Station
Detroit, Mich. 48219
*Aluminum Curtain Walls*
Architectural Aluminum Manufacturers Association
410 N. Michigan Avenue
Chicago, Ill. 60611
*Metal Curtain Wall Manual*
National Association of Architectural Metal Manufacturers
1010 West Lake Street
Oak Ridge, Ill. 60301
*Glazing Manual*
Flat Glass Marketing Association
1325 Topeka Avenue
Topeka, Kansas 66612
*Recommended Specifications for Lathing, Furring, and Plastering*
Contracting Plasters and Lathers International Association
20 E Street, N.W.
Washington, D.C. 20001
*Technical Notes—Brick and Tile*
Brick Institute of America
1750 Old Meadow Road
McLean, Va. 22101

## Manufacturers' Catalog Files

Manufacturers' catalogs represent another specification reference source. The suggested specifications in manufacturers' catalogs should be used with caution by specifiers. Whereas some publications include manufacturers' specifications which are accurately drawn, others are vague, and so written as to exclude certain items of work, and leave much to be desired in the way of precise, informative, and clear subject matter and specifications.

It is absolutely essential, when using specifications contained in manufacturers' catalogs, to be very discriminating in copying them verbatim. Do not use any clauses as written unless every statement is clearly understood. Modify the language where necessary to insure competition and complete understanding.

Many of these catalogs are contained in the following major publications of manufacturers' literature:

> *Sweets Architectural File*
> F. W. Dodge Corporation
> 1221 Avenue of the Americas
> New York, N.Y. 10020
> *A-E-C Western Catalog File*
> Times-Mirror Press
> 1115 South Boyle Street
> Los Angeles, Ca.

### General References

Additional reference sources for materials, workmanship, standards, tests, and general information are contained in the publications of various associations of manufacturers, technical societies, and contractors' associations. As an aid in quickly locating these sources of information, they have been arranged in accordance with the *Uniform Construction Index* as follows:

> *Division 1   General Requirements*
>
> *American Arbitration Association*
> 140 West 51 Street
> New York, N.Y. 10020
> *American Society of Safety Engineers, Inc.*
> 850 Busse Highway
> Park Ridge, Ill. 60065
> *Factory Insurance Association*
> 85 Woodland Street
> Hartford, Conn. 06102
> *Factory Mutual System*
> P.O. Box 688
> Norwood, Mass. 02062

## Division 2    Site Work

*American Association of Nurserymen*
Suite 835, Southern Building
Washington, D.C. 20005
*American Concrete Pipe Association*
1501 Wilson Boulevard
Arlington, Va. 22209
*Asphalt Institute*
1901 Pennsylvania Avenue, N.W.
Washington, D.C. 20006
*Chain Link Manufacturers Institute*
P.O. Box 515
Bronxville, N.Y. 10708
*Highway Research Board*
2101 Constitution Avenue, N.W.
Washington, D.C. 20418

## Division 3    Concrete

*American Concrete Institute*
Box 4754, Redford Station
Detroit, Mich. 48219
*Concrete Industry Board, Inc.*
51 East 42 Street
New York, N.Y. 10017
*Concrete Reinforcing Steel Institute*
228 North LaSalle Street
Chicago, Ill. 60601
*Expanded Shale, Clay, and Slate Institute*
1041 National Press Building
Washington, D.C. 20004
*Gypsum Roof Deck Foundation*
1201 Waukegan Road
Glenview, Ill. 60025
*Mo-Sai Institute, Inc.*
110 Social Hall Avenue
Salt Lake City, Utah

*National Ready Mixed Concrete Association*
900 Spring Street
Silver Spring, Md. 20910
*National Sand and Gravel Association*
900 Spring Street
Silver Spring, Md. 20910
*Portland Cement Association*
Old Orchard Road
Skokie, Ill. 60076
*Prestressed Concrete Institute*
205 West Wacker Drive
Chicago, Ill. 60606
*Vermiculite Association, Inc.*
527 Madison Avenue
New York, N.Y. 10022

## Division 4   Masonry

*The Barre Guild (Granite)*
51 Church Street
Barre, Vt. 05641
*Building Stone Institute*
420 Lexington Avenue
New York, N.Y. 10017
*Facing Tile Institute*
333 North Michigan Avenue
Chicago, Ill. 60601
*Indiana Limestone Institute of America, Inc.*
P.O. Box 489
Bloomington, Ind. 47401
   *Marble Institute of America, Inc.*
Pennsylvania Building, Room 848
425 13th Street, N.W.
Washington, D.C. 20004
*Masonry Institute of America*
2550 Beverly Boulevard
Los Angeles, Ca. 90057

*National Concrete Masonry Association*
2009 14th Street North
Arlington, Va. 22201
*Brick Institute of America*
1750 Old Meadow Road
McLean, Va. 22101


*Division 5   Metals*

*Aluminum Association*
750 Third Avenue
New York, N.Y. 10017
*American Hot Dip Galvanizers Association*
1000 Vermont Avenue, N.W.
Washington, D.C. 20005
*American Institute of Steel Construction, Inc.*
101 Park Avenue
New York, N.Y. 10017
*American Iron and Steel Institute*
150 East 42nd Street
New York, N.Y. 10017
*American Welding Society*
345 East 47 Street
New York, N.Y. 10017
*Architectural Aluminum Manufacturers Association*
410 N. Michigan Avenue
Chicago, Ill. 60601
*Lead Industries Association, Inc.*
292 Madison Avenue
New York, N.Y. 10017
*Porcelain Enamel Institute, Inc.*
1900 L Street, N.W.
Washington, D.C. 20036
*Steel Deck Institute*
9836 West Roosevelt Road
Westchester, Ill. 60153

*Steel Joist Institute*
2001 Jefferson Davis Highway
Suite 707
Arlington, Va. 22202
*Zinc Institute, Inc.*
292 Madison Avenue
New York, N.Y. 10017


## Division 6    Wood and Plastics

*American Forest Products Industries, Inc.*
1816 N Street, N.W.
Washington, D.C. 20036
*American Hardwood Association*
20 North Wacker Drive
Suite 1452
Chicago, Ill. 60606
*American Institute of Timber Construction*
1100 17th Street, N.W.
Washington, D.C. 20036
*American Plywood Association*
1119 A Street
Tacoma, Wash. 98401
*American Walnut Manufacturers Association*
666 North Lake Shore Drive
Chicago, Ill. 60611
*American Wood-Preservers Association*
2600 Virginia Avenue, N.W.
Washington, D.C. 20037
*Appalachian Hardwood Manufacturers, Inc.*
414 Walnut Street
Cincinnati, Ohio 45202
*Architectural Woodwork Institute*
Chesterfield House, Suite A
5055 S. Chesterfield Road
Arlington, Va. 22206

*California Redwood Association*
617 Montgomery Street
San Francisco, Ca. 94111
*Pine Hardwoods Association*
666 Lake Shore Drive
Chicago, Ill. 60611
*Forest Products Research Society*
2801 Marshall Court
Madison, Wisc. 53705
*Hardwood Dimension Manufacturers Association*
3813 Hillsboro Road
Nashville, Tenn. 37215
*Hardwood Plywood Manufacturers Association*
2310 South Walter Reed Drive
Box 6246
Arlington, Va. 22206
*National Forest Products Association*
1619 Massachusetts Avenue, N.W.
Washington, D.C. 20036
*National Hardwood Lumber Association*
59 East Van Buren Street
Chicago, Ill. 60605
*National Oak Flooring Manufacturing Association*
814 Sterick Building
Memphis, Tenn. 38103
*National Particleboard Association*
711 14th Street, N.W.
Washington, D.C. 20005
*National Woodwork Manufacturers Association, Inc.*
400 West Madison Street
Chicago, Ill. 60606
*Northern Hardwood and Pine Manufacturers Association*
Suite 207, Northern Building
Green Bay, Wisc. 54301
*Philippine Mahogany Association, Inc.*
P.O. Box 279
South Pasadena, Ca.

*Ponderosa Pine Woodwork Association*
39 South LaSalle Street
Chicago, Ill. 60603
*Red Cedar Shingle and Handsplit Shake Bureau*
5510 White Building
Seattle, Wash. 98101
*Southern Forest Products Association*
P.O. Box 52468
New Orleans, La. 70150
*Southern Pine Association*
520 National Bank of Commerce Building
Box 52468
New Orleans, La. 70150
*Vacuum Wood Preservers Institute*
5151 Holmes Road
Houston, Texas 77033
*West Coast Lumber Inspection Bureau*
P.O. Box 25406
Portland, Ore. 97225
*Western Red Cedar Lumber Association*
Yeon Building
Portland, Ore. 97204
*Western Red and Northern White Cedar Association*
Box 2576
New Brighton, Minn. 55112
*Woodworking Institute of California*
1417 Georgia Street
Los Angeles, Ca. 90015


*Division 7     Thermal and Moisture Protection*

*Adhesive and Sealant Council*
1410 Higgins Road
Park Ridge, Ill. 60068
*Asphalt Institute*
1901 Pennsylvania Avenue, N.W.
Washington, D.C. 20006

*Asphalt Roofing Manufacturing Association*
757 Third Avenue, Room 2111
New York, N.Y. 10017
*Building Waterproofer's Association*
60 East 42nd Street
New York, N.Y. 10017
*Copper Development Association*
405 Lexington Avenue
New York, N.Y. 10017
*National Mineral Wool Insulation Association*
211 East 51 Street
New York, N.Y. 10022
*National Roofing Contractors Association*
1515 North Harlem Avenue
Oak Park, Ill. 60302


*Division 8    Doors and Windows*

*American Society of Architectural Hardware Consultants*
77 Mark Drive
P.O. Box 3476
San Rafael, Ca. 94902
*Builders Hardware Manufacturers' Association*
60 East 42nd Street
New York, N.Y. 10017
*Door Operator and Remote Controls Manufacturers Association*
110 North Wacker Drive
Chicago, Ill. 60606
*Flat Glass Marketing Association*
1325 Topeka Avenue
Topeka, Kansas 66612
*National Association of Architectural Metal Manufacturers*
1010 West Lake Street
Oak Ridge, Ill. 60301
*National Builders Hardware Association*
1290 Avenue of the Americas
New York, N.Y. 10019

*Sealed Insulating Glass Manufacturing Association*
200 South Cook Street, Suite 209
Barrington, Ill. 60010
*Stained Glass Association of America*
3600 University Drive
Fairfax, Va. 22030
*Steel Door Institute*
2130 Keith Building
Cleveland, Ohio 44115
*Steel Window Institute*
2130 Keith Building
Cleveland, Ohio 44115


## Division 9    Finishes

*Acoustical and Insulating Materials Association*
205 West Touhy Avenue
Park Ridge, Ill. 60068
*Asphalt and Vinyl Asbestos Tile Institute*
101 Park Avenue
New York, N.Y. 10017
*Canvas Awning Institute*
1918 North Parkway
Memphis, Tenn. 38112
*Ceramic Tile Institute*
700 N. Virgil Avenue
Los Angeles, Ca. 90029
*Contracting Plasterers and Lathers International Association*
20 E Street, N.W., Suite One
Washington, D.C. 20001
*Gypsum Association*
201 North Wells Street
Chicago, Ill. 60606
*Gypsum Drywall Contractors International*
127 South Wacker Drive, Suite 201
Chicago, Ill. 60606

*Maple Flooring Manufacturers Association*
424 Washington Avenue
Oshkosh, Wisc. 54901
*Metal Lath Association*
636 Engineers Building
Cleveland, Ohio 44114
*National Lime Association*
4000 Brandywine Street, N.W.
Washington, D.C. 20016
*National Oak Flooring Manufacturing Association*
814 Sterick Building
Memphis, Tenn. 38103
*National Paint and Coatings Association*
1500 Rhode Island Avenue, N.W.
Washington, D.C. 20005
*National Terrazzo and Mosaic Association, Inc.*
716 Church Street
Alexandria, Va. 22314
*Paint and Wallpaper Association of America, Inc.*
2101 South Brentwood Boulevard
St. Louis, Mo. 63144
*Painting and Decorating Contractors of America*
2625 West Peterson Avenue
Chicago, Ill. 60645
*Perlite Institution*
45 West 45 Street
New York, N.Y. 10036
*Steel Structures Painting Council*
4400 Fifth Avenue
Pittsburgh, Pa. 15213
*Tile Council of America, Inc.*
800 Second Avenue
New York, N.Y. 10017
*Vermiculite Association, Inc.*
527 Madison Avenue
New York, N.Y. 10022

*Wall Coverings Council, Inc.*
969 Third Avenue
New York, N.Y. 10022
*Wood Flooring Institute of America*
201 North Wells Street
Chicago, Ill. 60606

### Division 10 Specialties

*Gas Appliance Manufacturers Association, Inc.*
60 East 42 Street
New York, N.Y. 10017
*National Association of Mirror Manufacturers*
1225 19th Street, N.W.
Washington, D.C. 20036
*National Kitchen Cabinet Association*
918 Commonwealth Building
Louisville, Ky. 40202

### Division 11 Equipment

*Association of Home Appliance Manufacturing*
20 North Wacker Drive
Chicago, Ill. 60606

### Division 12 Furnishings

*Carpet and Rug Institute*
208 West Cuyler Street
Dalton, Ga. 30720

### Division 13 Special Construction

*Incinerator Institute of America*
1 Stone Plaza
Bronxville, N.Y.
*National Swimming Pool Institute*
2000 K Street, N.W.
Washington, D.C. 20006

## Division 14    Conveying Systems

*International Material Handling Society*
2200 Fuller Road
Ann Arbor, Mich. 48105
*Materials Handling Institute, Inc.*
Gateway Towers, Gateway Center
Pittsburgh, Pa. 15222
*National Elevator Manufacturing Industry*
101 Park Avenue
New York, N.Y. 10017

## Division 15    Mechanical

*Air Conditioning and Refrigeration Institute*
1815 North Fort Myer Drive
Arlington, Va. 22209
*American Pipe Fittings Association*
60 East 42 Street
New York, N.Y. 10017
*American Society of Heating, Refrigerating, and
   Air Conditioning Engineers*
Engineering Center
345 East 47 Street
New York, N.Y. 10017
*American Water Works Association, Inc.*
2 Park Avenue
New York, N.Y. 10016
*Cast Iron Pipe Research Association*
1211 West 22 Street
Oak Brook, Ill. 60521
*Cast Iron Soil Pipe Institute*
1824–26 Jefferson Place, N.W.
2029 K Street, N.W.
Washington, D.C. 20006
*Cooling Tower Institute*
4242 Richmond Avenue
Houston, Texas 77027

*Home Ventilating Institute*
1108 Standard Building
Cleveland, Ohio 44113
*National Clay Pipe Institute*
69 North Wren Drive
Pittsburgh, Pa. 15243

## Division 16    Electrical

*American Home Lighting Institute, Inc.*
360 North Michigan Avenue
Chicago, Ill. 60601
*Illuminating Engineering Research Institute*
345 East 47 Street
New York, N.Y. 10017
*Illuminating Engineering Society*
345 East 47 Street
New York, N.Y. 10017
*Lightning Protection Institute*
2 North Riverside Plaza
Chicago, Ill. 60606

# 16

## Materials Evaluation

There are many factors to be considered when selecting and evaluating materials or products to be used in construction. It is imperative that the specifier have a full understanding of the problem confronting him. Partial or incomplete assessment can cause problems or failures since all possibilities have not been evaluated. It is therefore essential to state the problem and then to determine all its parameters.

At the outset, define the problem in terms of its needs to satisfy the design requirements. The needs or major considerations to be investigated can be listed as follows:

1. Function
2. Aesthetics
3. Serviceability and environment
4. Compatibility
5. Construction demands
6. Code requirements
7. Economics
8. Maintenance

The initial determination in the evaluation of a material must of necessity deal with function. For function, materials are evaluated on the basis of sound reduction, thermal efficiency, fire safety, weatherproofing, and other similar requirements. Parameters for each function must then be established and investigated.

Sound reduction may involve sound absorption and/or reduction in sound transmission. For sound absorption, the criterion is the noise reduction coefficient (NRC). The NRC should be determined for the space involved and then various materials can be reviewed for this rating from which to make the selection. If reduction in sound transmission is essential, the materials or composite constructions are reviewed with respect to the Sound Transmission Class (STC), ASTM E 90.

When fire protection is essential, the parameters for the building material or product to be checked against include flame spread, ASTM E 84 (tunnel test), or combustibility, ASTM E 136, or hourly rating, ASTM E 119.

When thermal efficiency or heat gain or loss is required, the parameter that the material or product must be equated against is its $K$ factor or conductivity which is the time rate at which heat flows through a homogeneous material 1 in. thick by 1 $ft^2$ in 1 hr when the temperature difference is 1°F.

Weatherproofing can involve water infiltration, moisture migration, vapor transmission, or other similar conditions. In each instance the parameters must be determined and the materials in question investigated to meet the criteria established. For vapor transmission, for example, a specific perm rating might be required and the material would be judged against this requirement.

However, after an investigation is made into the functional aspects of the material, the specifier must then take into account the other needs or design requirements. Not all materials require an evaluation concerning its aesthetics, especially if they are hidden or not exposed in the final construction. However, a sealant that is visual requires an assessment as to whether specific colors will not cause problems. A paint or coating with special requirements for color, texture, or gloss can be affected in so far as its life potential is concerned since reformulation to meet these requirements can alter the physical properties.

Serviceability and environment are the next variables to be investigated since these determine the durability of the materials under consideration. Serviceability is dependent on its use, whether it is an interior material subjected to people usage or an exterior material subjected to the environment. The service life within a structure subjects materials to the hazards of people in the form of wear, abrasion, vandalism, or rough usage. Service life within an industrial plant exposes materials to moisture, chemical interaction, industrial fumes, impact, fork lift trucks, skids, and so on. Environmental hazards to which exterior materials are subjected is their relationship to weather conditions. Weather factors that interact with materials are water, ozone, ultraviolet light, wind pressures, hailstones, snow, ice, temperature variation, and combinations of these elements. Again, the degree to which the material or product is subjected must be established to ascertain its performance.

For example, a function involving sound control in an enclosed swimming pool area would require the use of an acoustical material. However, the serviceability factor to be considered is the significant amount of moisture present in the space. The durability of the acoustical material selected to withstand the effect of moisture is of prime importance. A gypsum or wood fiber product would be susceptible to swelling and subsequent damage. A ferrous metal suspension system for the acoustical material would be subject to corrosion. The search for an acceptable system to endure the environmental atmosphere which is a measure of its service life would be narrowed to the selection of a corrosion resistant and moisture resistant material.

After the functional, aesthetic, and serviceability attributes are determined, the specifier must then investigate whether the material under consideration will be compatible with other materials that are combined with one another in a specific detail. The compatibility or ability of several materials to function together without deterioration, degradation, or interaction must be investigated. Generally, the chemical composition of each of the materials in a composite design requires investigation to forestall or minimize the incompatibility. For example, a sealant in an expansion joint must be equated against the filler material and the substrates forming the joint to insure compatibility since incompatibility can result in staining the substrate, reversion of the sealant, adherence of the sealant to the filler in the joint, or other symptoms of degradation.

The next step is to take into account the problems inherent in construction demands, such as hazards, procedures, and sequences. A construction hazard, for example, is the location of a materials hoist alongside an exterior facade which can result in the deposition of corrosive materials such as concrete or plaster spattering. Erection sequences can cause delay in the application of covering materials thereby necessitating special precautions or protective measures.

In selecting materials for waterproofing, the specifier must provide a method of protection that will ensure the integrity of the waterproofing during the construction phase. Recognizing that construction will continue on top of waterproofing before final surfacing materials are placed over it, the specifier will require the introduction of a protective covering such as concrete topping, impregnated board, or other materials that will be permanently incorporated in the construction. It is therefore essential to review the selection of a material, determine whether the construction process will have an adverse effect on its properties, and build in certain safeguards if this possibility exists.

Code requirements impose additional constraints that must be investigated to determine whether a material will meet these requirements. Codes impose requirements for fire safety, health, noise abatement, strength, and so on. A material required for a specific function must, after being reviewed for the preceding design requirements, meet the additional requirements imposed on it due to the law of the place of building, namely, the local building codes.

Economics also play a very important role in the evaluation and selection of material. The proposed structure as determined by the building owner will be assumed to have a specific service life of 20, 40, or 50 years. When judgments are made for specific materials, one takes into account the life span of the structure and determines which of the materials will meet the lowest acceptable time elements and be able to perform its selected functions.

The final attribute to be considered is the ease of maintenance of a specific finish material. Some materials may be inexpensive in terms of initial costs. However, a building owner may very well desire a maintenance free, or relatively maintenance free, surface and will equate its initial installation plus maintenance cost over its long-term investment and decide in favor of a relatively expensive one-time installation cost.

## New Products

For new products there are two major areas that involve materials evaluation. The first deals with the development of a product or a material to fit a particular situation created by specific requirements. The second involves an evaluation of the properties of a newly developed material or product to determine if the manufacturer's claims match his test results, thus warranting the use of his product.

For a product to be developed to meet a specific requirement, the specifier must establish the conditions under which it is to be used and the criteria for testing and acceptance. For example, if a floor is to be subjected to unusual hazards, such as moisture, acid spillage, hot jet fuels, and printers' ink, a standard flooring material might not be available to satisfy all the design conditions. The specifier would have to establish the design criteria. He would have to determine which unusual fluids would be likely to spill on the floor and to what extent the proposed flooring should resist the effects of such spillage. He would have to take into account resistance to abrasion, slip resistance, indentation, hardness, heat resistance, and similar factors. He can establish the parameters by selecting certain ASTM test procedures by which these characteristics would be measured. After he determines which test procedure to use, he can set minimum and maximum values for the test results and ask manufacturers to formulate a product to meet these criteria. The end product by a manufacturer could be an epoxy, neoprene, polyester, acrylic, or a urethane formulation. The specific basic ingredient is not important to the architect and specifier. The end result or performance characteristics determined by the materials evaluation is all that is essential.

New products are developed by manufacturers either to fill a specific need or to improve existing products. For the most part, manufacturers have been taking the lead in developing new products rather than architects. After they are developed, the manufacturer proceeds to bring the items to the attention of architects and specifiers. Where the products are referenced by the manufacturer to a reference standard, such as a federal specification, ASTM, or ANSI specification, there is no major problem involved with evaluating the new product. However, many new products are specifically designed by the manufacturers to keep ahead of their competition. In these cases, the physical and chemical properties are not referenced to known standards. A specifier investigating these products finds them difficult to evaluate without normal standards of comparison. Sometimes the manufacturer develops his own test methods, and the results have no correlation with standard test procedures.

What procedures does a specifier follow in evaluating new products? He must take several factors into account. The integrity of the manufacturer. Has he had a successful record in the past for developing good products? Has he field tested the new product? Is there any correlation between his field tests and his laboratory testing? Has he tested the significant properties of the product?

The reliability of the source of information and its authenticity should be investigated. Check with other architects and engineers if they are given as references to determine whether the condition of use is similar to that proposed for your project. Demand additional test data if necessary. Suggest specific properties to be tested.

Review the problems to be encountered in the field in the handling and installation of a new product. Will there be an adequate fully trained corps of trades who understand how to handle the new product? Are there franchised applicators? Are there any special precautions to be observed with respect to volatile solvents, flammable materials, or staining of adjacent surfaces?

The evaluation of new or untried materials for possible use should include discussions with the manufacturer to obtain long-term guarantees to insure additional safeguards for the client and the design professional.

# 17

## Specifications Writing Procedures

How does one write a specification? The uninitiated practitioner faced with the task of writing a specification for his first project does what all other beginners have done who have not had a basic understanding of the principles of specification writing. In his emergency, he begs from some friend of older practice the specifications of another undertaking as like in character to his own as he can find, and then cuts, pastes, writes in, and crosses out as well as he knows how, to make a patchwork that will apply more or less to the structure he has planned.

However, armed with the principles of specification writing, the task becomes less onerous and more manageable. A system of specification writing procedures should include the work preliminary to the actual writing of the specifications, the outline or preliminary specifications, the sources of information, the form and arrangement of the specifications, the actual writing of the specifications, and, finally, the reproduction and binding of the specifications. These procedures deal with time-tested methods such as the use of guide or master specifications, checklists, work sheets, and catalog files.

Reduced to their simplest form, specifications should be written according to an organized system. A good draftsman develops systematic methods of laying out his drawings. A good office has logical standards for indication of doors, windows, and the other countless elements of the drawings. Similarly, a specifier must have a system for the preparation of specifications, especially since they must be written after the drawings have progressed to a point where they are about 50% completed and the time available to write and complete the specifications is scant. The pressure of time thus makes a systematic approach essential.

One of the first documents that the specification writer should have is preliminary, or outline, specifications. This is generally prepared by the project architect or designer with the collaboration of the specifier, and briefly lists materials and finishes without describing workmanship or fabrication. The next step is to prepare a complete takeoff of every item from the working drawings and—in conjunction with a standard checklist and the *Uniform Construction Index* described in Chapter 19—establish the technical sections.

With the technical sections established on the basis of the preliminary takeoff, the specifier is now in a position to start and complete some sections, and to start and gather information on other sections, or do research on some materials where he does not have sufficient information. The nature of specification writing is such that one cannot start writing immediately and continue until the project is completed. There will be need for conferences with the job captain and designer to arrive at decisions on many items, and it will be necessary to obtain information from manufacturers and their representatives on materials and products when the architectural details involving these items are in doubt and require clarification and research.

There are many sections that can be written on the basis of incomplete drawings. These should be written at the outset since they are not likely to change during the development of the drawings. Such sections include those under Division 9, Finishes, for example, ceramic tile, terrazzo, resilient flooring, and acoustic treatment. Other sections that can be written around partially completed drawings include earthwork, concrete, toilet partitions, and masonry.

To write these sections, many specifiers will have their own guide or master specifications, which they have carefully developed over the years. To be truly effective, these guides should not be static, and they should be revised as dictated by experience and new developments. Some people refer to these guides as canned specifications. However, it is difficult to see how any specification writer can do without such a valuable tool, which comprises the sum total of his experiences and his best efforts to write better specifications.

Specifiers, like any other individuals, naturally develop their own personal idiosyncrasies with respect to the systems they will develop in organizing themselves, their work habits, and their approach to the task of writing specifications. Some use card systems on which they develop standard paragraphs; others use collections of notes and checklists. Whatever system is employed, it should be orderly and systematic.

The following principles will aid the beginner in establishing a procedure for writing his specifications when he approaches the task before him.

1. Review the preliminary or outline specifications to obtain a better understanding of the project.

2. Review the preliminary drawings to visualize the project and obtain a better insight.

3. Since the architectural specifier is the focal point all the specifications, determine who the consultants are for the structural, mechanical, electrical, and site specifications. Coordinate their activities and establish the form, arrangement, and numbering system of the technical sections. To insure coordination between the respective sections so that there is no duplication or overlapping, submit a coordination list (see exhibit at the end of this chapter) to all the consultants for agreement on what goes where.

4. Review the working drawings and prepare a table of contents of the technical sections. (See Chapter 3 for typical section titles.)

5. Make a takeoff from the drawings of all the items and list them on work sheets under the appropriate section titles. For example, under the section title "Miscellaneous Iron and Steel," make a listing of such items as railings, ladders, stairs, saddles, gratings, and mesh partitions, and indicate the drawings on which the details occur so that they can be easily found again when the final specification is written.

6. Discuss questions relating to any of these items with the job captain, designer, or any other individual, and determine what will be shown on the drawings and what will be specified (see Chapter 2). Determine which items require additional research, note these, and

perform the necessary investigation at a time when a lack of sufficient drawings precludes actual writing of specifications.

7. Commence the actual writing of the specifications. Use guide or master specifications where these are available, and utilize the takeoff list and a checklist to insure completeness of each section.

8. Select those sections that will not be affected by further development of the drawings as previously described and complete these sections. Start those sections on which there is a good deal of information that can be gleaned from the drawings. Note the information that will be required in order to complete them at a later date. Arrange the information within each section as described in Chapter 5.

9. Do the required research on unknowns when you can no longer proceed with any actual specification writing.

10. Leave until the very last those sections that require almost complete working drawings, such as carpentry and millwork, and miscellaneous and ornamental metal.

SPECIFICATION COORDINATION
CHECKLIST
PROJECT NO. _____

Project _____

Project Architect _____

Mechanical and Electrical
Consultants _____

Structural Consultants _____

Site Consultants _____

Architectural   Food Service   Conveying Systems   Plumbing   Heating and Ventilation   Sprinkler   Electrical

Mark as follows:
F. Furnish   I. Install   FI. Furnish and Install

Temporary water .....................
Temporary toilets ....................
Temporary heat ......................
Temporary fire protection .............
Temporary light and power ...........
Temporary emergency lighting ........
Excavation and backfill inside building
    for each trade if not by
    Architectural .......................

| | SPECIFICATION COORDINATION CHECKLIST | Architectural | Food Service | Conveying Systems | Plumbing | Heating and Ventilation | Sprinkler | Electrical |
|---|---|---|---|---|---|---|---|---|
| 8 | Excavation and backfill outside building for each trade if not by Architectural .................... | | | | | | | |
| 9 | Keeping site and excavation free from water ........................... | | | | | | | |
| 10 | Underfloor drains ................... | | | | | | | |
| 11 | Footing drains ...................... | | | | | | | |
| 12 | Drywalls ........................... | | | | | | | |
| 13 | Connection of underfloor and footing drains to storm drain system ........ | | | | | | | |
| 14 | Forms for foundations and pads for trade items .................... | | | | | | | |
| 15 | Concrete for foundations and pads ............................. | | | | | | | |
| 16 | Headwalls .......................... | | | | | | | |
| 17 | Septic tank ......................... | | | | | | | |
| 18 | Disposal field ....................... | | | | | | | |
| 19 | Drainage manholes ................... | | | | | | | |
| 20 | Sanitary manholes ................... | | | | | | | |
| 21 | Steam manholes ..................... | | | | | | | |
| 22 | Electrical manholes and handholes ........................ | | | | | | | |
| 23 | Drainage catch basins ................ | | | | | | | |
| 24 | Drainage manhole frames and covers ............................ | | | | | | | |
| 25 | Sanitary manhole frames and covers ............................ | | | | | | | |
| 26 | Steam manhole frames and covers ....... | | | | | | | |
| 27 | Electrical manhole and handhole frames and covers ................. | | | | | | | |
| 28 | Drainage catch basin frames and covers ............................ | | | | | | | |

| SPECIFICATION COORDINATION CHECKLIST | Architectural | Food Service | Conveying Systems | Plumbing | Heating and Ventilation | Sprinkler | Electrical |
|---|---|---|---|---|---|---|---|
| Pit frames and covers ................. | | | | | | | |
| Catwalks to trade equipment ......... | | | | | | | |
| Ladders to trade equipment and valves ............................. | | | | | | | |
| Supplementary steel for trade equipment ....................... | | | | | | | |
| Ornamental HVAC grilles ........... | | | | | | | |
| Exterior wall louvers ................ | | | | | | | |
| Louver connections to ducts .......... | | | | | | | |
| Vent pipe cap flashing ............... | | | | | | | |
| Vent pipe base flashing .............. | | | | | | | |
| Curb cap flashing for trade equipment ....................... | | | | | | | |
| Curb base flashing for trade equipment ....................... | | | | | | | |
| Roof drains ........................ | | | | | | | |
| Roof drain flashing ................. | | | | | | | |
| Waterproofing of ceramic tile showers ........................... | | | | | | | |
| Waterproofing of mop receptors ........ | | | | | | | |
| Thermal insulation of boiler room ceiling ........................... | | | | | | | |
| Door louvers ....................... | | | | | | | |
| Access panels and support frames in plaster ........................ | | | | | | | |
| Access panels and support frames in masonry ....................... | | | | | | | |
| Access panels and support frames in acoustical tile ..................... | | | | | | | |
| Vermiculite fireproof covering ......... | | | | | | | |
| Cutting to accommodate trade items ............................ | | | | | | | |
| Rough patching .................... | | | | | | | |

| SPECIFICATION COORDINATION CHECKLIST | Architectural | Food Service | Conveying Systems | Plumbing | Heating and Ventilation | Sprinkler | Electrical |
|---|---|---|---|---|---|---|---|
| 52 Finish patching ..................... | | | | | | | |
| 53 Prime painting of trade piping and ductwork ..................... | | | | | | | |
| 54 Finish painting of trade piping and ductwork ..................... | | | | | | | |
| 55 Prime painting of trade equipment ......................... | | | | | | | |
| 56 Finish painting of trade equipment ......................... | | | | | | | |
| 57 Color coding of piping .............. | | | | | | | |
| 58 Toilet room accessories .............. | | | | | | | |
| 59 Central soap dispensing system ........ | | | | | | | |
| 60 Kitchenette unit ..................... | | | | | | | |
| 61 Kitchenette unit connections .......... | | | | | | | |
| 62 Walk-in refrigerator ................. | | | | | | | |
| 63 Walk-in refrigerator compressors, piping, and controls ..................... | | | | | | | |
| 64 Valved supplies, waste, and vent piping for food service equipment ......................... | | | | | | | |
| 65 Valved supplies, waste, and vent piping for laboratory equipment ......................... | | | | | | | |
| 66 Valved supplies, waste, and vent piping for hospital equipment ....... | | | | | | | |
| 67 Sink strainers and tailpieces ........... | | | | | | | |
| 68 Supply fittings for food service equipment ......................... | | | | | | | |
| 69 Supply fittings for laboratory equipment ......................... | | | | | | | |
| 70 Supply fittings for hospital equipment ......................... | | | | | | | |
| 71 Greasetraps ......................... | | | | | | | |

| SPECIFICATION COORDINATION CHECKLIST | Architectural | Food Service | Conveying Systems | Plumbing | Heating and Ventilation | Sprinkler | Electrical |
|---|---|---|---|---|---|---|---|
| Booster heaters ..................... | | | | | | | |
| Hoods for food service equipment ........................ | | | | | | | |
| Hoods for dishwashing equipment ....................... | | | | | | | |
| Hoods for laboratory equipment ....... | | | | | | | |
| Ductwork for hoods ................. | | | | | | | |
| X-ray equipment ................... | | | | | | | |
| X-ray equipment supports ............ | | | | | | | |
| X-ray equipment connections ......... | | | | | | | |
| Sterilizer equipment ................. | | | | | | | |
| Sterilizer equipment connections ....... | | | | | | | |
| Surgical lights ...................... | | | | | | | |
| Surgical light supports ............... | | | | | | | |
| Surgical light connections ............ | | | | | | | |
| X-ray film illuminators .............. | | | | | | | |
| X-ray film illuminator connections ....................... | | | | | | | |
| Panel heating system ................ | | | | | | | |
| Panel heating system connections ....... | | | | | | | |
| Sound control rooms ................. | | | | | | | |
| Sound control room silencers .......... | | | | | | | |
| Sound control room wiring and devices ...................... | | | | | | | |
| Sound control room connections ....... | | | | | | | |
| Elevator machine support beams ....... | | | | | | | |
| Elevator hoistway frames, doors, and saddles ................. | | | | | | | |
| Conveying system controls ........... | | | | | | | |
| Conveying system disconnect switch and power wiring ................. | | | | | | | |
| Moving stair and sidewalk frames ...... | | | | | | | |

| SPECIFICATION COORDINATION CHECKLIST | Architectural | Food Service | Conveying Systems | Plumbing | Heating and Ventilation | Sprinkler | Electrical |
|---|---|---|---|---|---|---|---|
| 98 Sidewalk elevator frame and door ...... | | | | | | | |
| 99 Linen and garbage chutes ............. | | | | | | | |
| 100 Window washing equipment .......... | | | | | | | |
| 101 Pneumatic tube system ............... | | | | | | | |
| 102 Conveyor system ..................... | | | | | | | |
| 103 Shower stall pan flashing ............. | | | | | | | |
| 104 Gang showers ....................... | | | | | | | |
| 105 Prefabricated showers ................ | | | | | | | |
| 106 Fire hose and extinguisher cabinets ........................... | | | | | | | |
| 107 Fire extinguishers .................... | | | | | | | |
| 108 Incinerator .......................... | | | | | | | |
| 109 Incinerator connections ............... | | | | | | | |
| 110 Chimney breeching frame ............. | | | | | | | |
| 111 Chimney cleanout door .............. | | | | | | | |
| 112 Prefabricated chimney ............... | | | | | | | |
| 113 Convector enclosures ................ | | | | | | | |
| 114 Fan coil enclosures .................. | | | | | | | |
| 115 Induction unit enclosures ............. | | | | | | | |
| 116 Cabinet heater enclosures ............. | | | | | | | |
| 117 Lighting fixture supports ............. | | | | | | | |
| 118 Plaster rings for lighting fixtures ....... | | | | | | | |
| 119 Lightning protection ................. | | | | | | | |
| 120 Watchmen's system .................. | | | | | | | |
| 121 Intercommunications system .......... | | | | | | | |
| 122 Clocks ............................. | | | | | | | |
| 123 Exterior transformer vault ........... | | | | | | | |
| 124 Transformer vault entrance .......... | | | | | | | |

# 18

## Computerized Specifications

The application of the computer to the development of project specifications was begun seriously in 1966. There was a compelling need to automate specifications as a result of the information and materials explosion. With thousands of new materials and building products available to the architect and specifiers, it became apparent that it was essential to reduce this accumulation of invaluable information to some manageable form. This was necessary so that the architect, engineer, and specifier could concentrate on design, research, and problem solving rather than in cutting and pasting together a project specification.

Fortunately, a tool was available to cope with the problem. The hardware was the computer. However, the task at hand was to develop a software program that would manipulate language. Software in computer parlance is defined as programs and accessory components. At the outset, the problem was magnified because the specifier and the computer expert could not communicate one another's expertise in a common language. The specifier was aware of the complexity of specifications and the computer expert was proficient in computer hardware and software components. What was essential was a bridging of this communications gap.

The first step toward automation began by employing the automatic typewriter such as the IBM MT/ST and the Dura typewriter. With the recognition that automatic typewriters and computers could solve the problem of preparing project specifications other than by cutting and pasting came the next major consideration, namely, that in order for these devices to be successful it was essential to create master specifications. However, this feature, the master specification, was abhorrent to many specifiers.

For years, many specifiers avoided the use of master specifications. The concept of a master specifications did not sit well with some specifiers who viewed masters in the light of that odious connotation canned specs. Those who think in these terms believe that each project specification is an original effort. In reality it is not so. Any new specification is developed from the last project specification that was issued, and at least 70 to 80% is usually reused with some editing. The only problem is that no one working without a master specification has ever taken the time to update and clean up an old specification. The same errors are repeated. With a master, a separate effort is required to organize, verify, update, and formulate a meaningful master specification geared to the practice and peculiarities of a specific office. Many specifiers were opposed to what they considered to be frozen, unchanging, and confining master specifications. Yet it was apparent that the automatic typewriter and the computer were inoperable without a carefully prepared master specification. However, when one recognized that the computer and the master specification are servants and tools of the specifier and that neither could supplant the specifier in his role as decision maker, the specifier accepted these new tools. Since these devices are essential to control the mass of information that the specifier must assess, it became apparent that he would have to accept these devices in order to cope with the problem which was getting out of hand.

The term "automated" is often used to describe the function of the computer in producing specifications, and some specifiers were concerned with the ultimate possibility of being displaced by these new devices. Since the term "specifier" does not fully describe the duties and responsibilities of this individual, one is inclined to assume that a device which takes over the laborious task of producing specifications must somehow automatically replace him. To the extent that one can anticipate future developments, it would appear that the talents of the specifier will be required far more as a materials researcher than at present, and far less as a writer since this task will be performed by a machine. The constant development of man-made products derived through chemistry and metallurgy is placing an increased demand on individuals who practice as specifiers, but who should be spending more time in materials engineering and research.

The advent of the computer can now free these professionals from the time-consuming task of cutting, clipping, writing, and pasting together a patchwork of specification paragraphs. The computer has arrived at a propitious moment to take over the task of producing a specification automatically to be used as a tool and not as a decision maker. The selection process is still the responsibility of the architect and specifier, and the computer spews out the results of their decisions.

To make the computer operational, two major ingredients are essential. First and foremost is a master specification, and secondly a software program is required that can manipulate the specification text. The ability to prepare master specifications for a computer is the key to its successful operation. Today, after about seven years of experience with computerized specifications, several national master specifications, tailored to the computer and its special needs, have evolved. Notably these masters are Production Systems for Architects and Engineers (PSAE), sponsored by AIA, and Pacific International Computing (PIC), a division of the Bechtel Corporation.

The central question is whether broad national master specifications can serve the differing needs of all offices. For the neophyte, and the small offices that either do not have the talent or the time to invest in producing their own master specifications, these national masters do provide a reasonably good starting point.

For the firm with experienced specifiers, the nationally available masters are rewritten by the specifier to reflect his experience, judgment, and the type of work performed by his firm. There is no one master specification that fits the needs of every office. A master must be custom designed by the specifier for the particular needs, decisions, and practices of his firm.

To prepare a master specification for a specific office, the specifier must take into account the types of projects generally undertaken by that firm: hospitals, schools, apartment houses, office buildings, shopping centers, or a combination of these building types. In addition, the designers in the firm might have preferences for the type of medium in which they design, such as masonry, concrete, or metal and glass for exterior facades. The specifier should then determine which technical sections would normally comprise the work included in these projects. He might find that he could very well limit the scope of his masters to about 50 or 60 technical sections that occur quite frequently. He could then establish his table of contents and set priorities for these master sections to concentrate on. Obviously, earthwork, concrete, miscellaneous metal, roofing, resilient flooring, painting, ceramic tile, and others can be used consistently from project to project.

The next priority might be masonry, drywall, waterproofing, dampproofing, terrazzo, and metal toilet partitions. A third priority would comprise carpentry, millwork, and cabinet work. If the office specialized in hospitals, technical sections on X-ray construction, conductive flooring, and special resin, matrix flooring would be undertaken early. If schools predominate, technical sections on chalkboards, tackboards, gymnasium equipment, and so on would be early priorities. In each instance it would be advisable to determine the frequency with which certain technical sections were used and to develop these first; those used less frequently would be assigned a third or fourth priority in their development.

Master specifications custom designed to specific offices should not be written to encompass every possible material or alternative situation since the time required to produce such a document would be endless and quite unnecessary. If one makes the assumption that he will prepare a master based on the frequency with which a material is used, he can reduce his sights considerably and produce a master that can be edited and expanded as the need arises.

A master specification cannot be economically and successfully prepared for every technical section. Some specialty and equipment items specified under Divisions 10 and 11 of the *Uniform Construction Index* can be difficult to formulate into a master specification. For example, if an office were involved primarily with the design of schools, the specifier, together with the designer, might select certain toilet accessories for inclusion in a master specification. However, in a large office with many designers with a wide practice in hospitals, office buildings, apartment houses, and other varied structures, the specifier could spend endless hours attempting to prepare a master for toilet accessories. The decision could be to write a project specification for toilet accessories as the need arises and not attempt to write a master. One must weigh the advantages and disadvantages of employing a computer with its master specifications and recognize that some specification sections, particularly for Divisions 10 and 11, might still lend themselves to easier preparation by the old cut and paste method.

In addition, as one develops a rapport between the capability of the computer and the software program on the one hand, and the master specification on the other, one recognizes that smaller technical sections, even more fragmented than those currently listed in the *Uniform Construction Index*, must be devised to arrive at a workable master.

For example, if the specifier were to prepare a master specification for a curtain wall, it would create quite a ponderous tome that would have to include all of the following: metal framing and panel systems in each of the following metals—aluminum, stainless steel, brass and bronze, and steel, and for all of these metals the specifier would have to provide information on all the finishes available. For the aluminum he would have natural anodized, color anodized, and baked enamel; for stainless steel he would specify the various polished finishes that are available; for brass and bronze he would include statuary finishes and various alloys for sheet and extrusions; for carbon steel he would have to include galvanizing and various priming and painting systems. And that only covers the metal portion. Then there would have to be included glass and glazing systems which are endless in the profusion of the different types of glass available—heat absorbing, heat reflecting, wire, clear, plate, float, insulating, ad infinitum.

Glazing systems would include sealants of all types, such as polysulfides, silicones, polyurethanes, butyls and glazing compounds, glazing gaskets, lockstrip gaskets, and so on. Insulation materials would include fiberglass, mineral wool, perlite, cellular glass, and their installation. Flashing materials of different types would include the metals such as copper, aluminum, stainless, and zinc, and the fabrics and elastomeric sheet systems. The inclusion of all the materials that could possibly comprise a curtain wall master specification would be awkward and unworkable. But, then, how does one get around this situation?

The curtain wall master could be broken down into a series of related narrow scope sections, each one dealing with a portion of the overall curtain wall. The first master in this series would be called "Curtain Wall General," in which the specifier would include the scope of the section, outline the qualifications of the subcontractor, set forth the criteria for the performance of the curtain wall against wind loads, thermal movement, air and water infiltration, and the installation of all component parts. A separate master would then be written for a mock-up and a testing program for the curtain wall. A separate master would be prepared for glass and glazing, for sealants, for thermal insulation, for flashing, for aluminum window wall components, for steel window wall components, and for other metal window wall components. Each of these sections would then be edited for specific projects and the section title names would change so that glass and glazing would read "Curtain Wall Flashing" and so on down the line so that there might be six, seven, or eight sections comprising the curtain wall for a specific project. Glass, sealants insulation, and flashing for other portions of the structure would be specified under headings of the same name with a notation under the scope of each of these to indicate that the materials and installation of these materials for the curtain wall are specified under the appropriate curtain wall sections.

As one gains familiarity with the manipulation of the computer, one would recognize that it is costly to make many editing changes. To reduce this cost it might prove more expedient to put some information on the drawings that might normally appear in the specifications. For example, the master specification for resilient flooring would contain the language that resilient tiles were $9 \times 9 \times 1/8$ in. unless otherwise shown, and the drawings would be annotated to indicate the location of sizes other than $9 \times 9$. Sealants, flashing, waterproofing, and thermal insulation would have to be shown on the drawings and called for in the specifications. We simply cannot indicate flashing, sealants, or waterproofing on the drawings. We must make further identification on that document so that we reduce the amount of editing in the computerized specification. The revolution that is now taking place for those specifiers who are utilizing or will be utilizing the computer for specifications will similarly take place on the drawing board, and somewhere along the line in this development, new guide lines can be formulated as to what should be drawn and what should be specified.

Another aspect of preparing master specifications is related to the "relational option" that has been conceived for the software computer program. This feature allows the specifier to tie together related text material within a technical section. This aspect requires a rewriting of specification technical sections so that the related material is not embedded within a paragraph, but is singled out so that it stands alone. See Exhibit 1 for original material and Exhibit 2 for the re-written material, illustrating how the relational option stands apart and is recognized and deleted by the computer. As a result of the relational option the specifier preparing master specifications must rewrite specification clauses and rearrange material and information so that related text material can be included or deleted automatically by the computer.

Initial software programs were not geared specifically for specification text manipulation. At the outset, specifiers commenced with software programs that were available to handle language. Nevertheless, this permitted experimentation and the development of ideas which led to the creation of superior software programs. Initially the software language computer program could perform the following tasks:

1. Edit text on a word, line, or paragraph basis.
2. Add text on a word, line, or paragraph basis.
3. Delete text on a word, line, or paragraph basis.
4. Provide headings and footings on each page for job numbers, section titles, and dates.
5. Number pages automatically.
6. Justify page widths, lengths, and margins.
7. Indent, underline, and space texts.

The shortcomings of the initial software programs became apparent to those who were using the early programs. Automated Procedures for Engineering Consultants (APEC), a group of engineers interested in the application of the computer toward solving engineering problems, established criteria for a specifications software program and developed a program known as SPECS. This program has the following features available to manipulate and edit specification text as follows:

1. Disappearing notes. Notes on master specification copy disappear on final project copy.
   a. Notes to specifiers—for options, coordination, and explanation.
   b. Notes to job captains—for drawing coordination and drawing notes.
2. Flexible formating. This feature is especially useful for engineers who must conform to architects' formats. This feature permits identification of articles, paragraphs, and subparagraphs by any alphanumeric arrangement and permits completely different vertical and horizontal arrangement known as a format file, that is, the text can appear in any format and can parallel the architect's arrangement very simply.
3. Automatic paragraph renumbering. Articles, paragraphs, and subparagraphs are automatically renumbered after addition or deletion of text.
4. Automatic generation of specific information. Separate information can be generated for separate lists of samples, shop drawings, guarantees, texts, certificates, and tables of contents. This is useful to the specifier, job captain, and field representative during construction and expedites search for this information in a voluminous specification.
5. Phrase options, Permits replacement of a word or phrase throughout the text. For example, if the architect does not have construction supervision, the word "architect" can be replaced by "owner," "contracting officer," or "inspector" by a single computer command.
6. Multiple choice blocks. This feature provides for the selection by the specifier of a choice among several options included in a master specification. Where one option is selected the others are automatically deleted.

7. Relational options. This feature, which was previously described, enables the specifier to delete articles, paragraphs, or sentences easily in a master specification that does not pertain to a specific project.

Exhibit 1 shows a page of a master specification. Exhibit 2 shows the same master with editing. Exhibit 3 shows the resultant project specification.

A master specification utilizing this specific software program can be altered for a specific project so that it becomes a submaster. This is accomplished by developing a questionnaire and a checklist. The questionnaire consists of the relational options and the checklist of the major article and paragraph headings. By reviewing a project with the job captain, utilizing the questionnaire and the checklist, the content of the master is reduced considerably. This step is taken initially to weed out the items from the master that are not required for a specific project. Exhibit 4 shows a typical questionnaire and checklist.

In addition to the APEC software program, AIAs PSAE has a computer program to handle its specification effort and CSI has developed COMSPEC as a software specification computer program.

34 4        1.  Submit certified copies of mill test reports for all
                steel furnished.  Perform mechanical and chemical
                tests for all material regardless of thickness or
                use.  No part of the ASTM specifications will be
                waived without written consent of the Architect.

                        **** OPTION    A-2     ****
35 4        2.  Submit certified statement from the base metal
                manufacturer that the proposed welding materials and
                techniques proposed for weathering steel will
                produce weldments meeting the specified requirements
                under actual project conditions.

36 2  1.4  PRODUCT HANDLING

37 3        a.  Do not handle structural steel until paint has thoroughly
                dried.  Care shall be exercised to avoid abrasions and
                other damage.

38 3        b.  Stack material out of mud and dirt and provide for proper
                drainage.  Protect from damage or soiling by adjacent
                construction operations.

                        **** OPTION    A-2     ****
39          c.  Weathering steel shall be stacked and/or handled in a
                manner which will prevent staining.

40 1                        PART 2 - PRODUCTS

41 2  2.1  MATERIALS

42 3        a.  Structural Steel: ASTM A 36 unless otherwise shown.

                        **** OPTION    A-2     ****
43 3        b.  Weathering Steel: ASTM A588 Grade A or B unless otherwise
                shown.  Use one (1) grade throughout.

                        **** OPTION    C      ****
44 3        c.  High-Strength Bolts and Nuts:  ASTM A325 and ASTM A490,
                minimum 3/4 in. diameter.

                        **** OPTION    B      ****
45 3        d.  Unfinished Bolts: ASTM A307 regular hexagon-bolt types,
                minimum 3/4 in. diameter.

                        **** OPTION    E      ****
46 3        e.  Rivets: ASTM A502.  Dimensions of rivets shall conform to
                requirements of "Handbook on Bolt, Nut and Rivet
                Standards", minimum 3/4 in. diameter.

                        5.1A-3                  Structural Steel

EXHIBIT 1    ORIGINAL UNEDITED MASTER SPECIFICATION

~~MASTER SPECIFICATION~~

34 4     1. Submit certified copies of mill test reports for all steel furnished. Perform mechanical and chemical tests for all material regardless of thickness or use. No part of the ASTM specifications will be waived without written consent of the Architect.

*Typist: Delete Option A-2*

             **** OPTION    A-2     ****

35 4     2. Submit certified statement from the base metal manufacturer that the proposed welding materials and techniques proposed for weathering steel will produce actual project conditions.

36 2 1.4 PRODUCT HANDLING

37 3     a. Do not handle structural steel until paint has thoroughly dried. Care shalll be exercised to avoid abrasions and other damage.

38 3     b. Stack material out of mud and dirt and provide for proper drainage. Protect from damage or soiling by adjacent construction operations.

             **** OPTION    A-2     ****

39     c. Weathering steel shall be stacked and/or handled in a manner ~~which~~ will prevent staining.

*stet*

40 1              PART 2 - PRODUCTS

41 2 2.1 MATERIALS

42 3     a. Structural Steel: ASTM A 36 *and A440 as shown.* ~~unless otherwise shown.~~

             **** OPTION    A-2     ****

43 3     b. Weathering Steel: ASTM A588 Grade A or B unless otherwise shown. Use onr (1) grade throughout.

             **** OPTION    C     ****

44 3     c. High-Strength Bolts and Nuts: ASTM A325 and ASTM A490 minimum 3/4 in. diameter.

             **** OPTION    B     ****

45 3     d. Unfinished Bolts: ASTM A307 regular hexagon-bolt types, minimum 3/4 in. diameter.

             **** OPTION    E     ****

46 3     e. Rivets: ASTM A502. Dimensions of rivets shall conform to requirements of "Handbook on Bolt, Nut and Rivet Standards", minimum ~~3/4~~ 1/2 in. diameter.

             5.1A-3                     Structural Steel

EXHIBIT 2     MASTER SPECIFICATION EDITED FOR PROJECT
PROJECT XYZ ANYWHERE, USA

1.  Submit certified copies of mill test reports for all
    steel furnished.  Perform mechanical and chemical
    tests for all material regardless of thickness or
    use.  No part of the ASTM Specifications will be
    waived without written consent of the Architect.

1.4 PRODUCT HANDLING

a.  Do not handle structural steel until paint has thoroughly
    dried.  Care shall be exercised to avoid abrasions and
    other damage.

b.  Stack material out of mud and dirt and provide for proper
    drainage.  Protect from damage or soiling by adjacent
    construction operations.

PART 2 - PRODUCTS

2.1 MATERIALS

a.  Structural Steel: ASTM A 36 and A 440 as shown.

b.  High-Strength Bolts and Nuts:  ASTM A325 and ASTM A490,
    minimum 3/4 in. diameter.

c.  Unfinished Bolts:  ASTM A307 regular hexagon-bolt types,
    minimum 3/4 in. diameter.

d.  Rivets:  ASTM A502.  Dimensions of rivets shall conform
    to requirements of "Handbook on Bolt, Nut and Rivet
    Standards", minimum 1/2 in. diameter.

5.1A-3                    Structural Steel

EXHIBIT 3     PROJECT SPECIFICATION COMPUTER PRINTOUT
PROJECT XYZ ANYWHERE, USA

MASTER SPECIFICATION QUESTIONAIRE

SECTION 5.1A

QUESTIONAIRE FOR STRUCTURAL STEEL

JOB NAME: _____

Strike out underlined letter-codes of items that DO NOT apply to
this job.

A Architectural Exposed Steel
    A-1 Non-Weathering
        A-1a Organic Paint (vs A-1b)
        A-1b Inorganic Paint (vs A-1a)
    A-2 Weathering

B Unfinished Bolts

C High Strength Bolts

D Welding (Mandatory for Architectural Exposed Steel)

E Riveting

F Stud Sheer Connectors
    F-1 Are Specified in this Section (vs F-2)
    F-2 Are Specified Elsewhere (vs F-1)

G Sliding Connections (for expansion joints)

H Elastomeric Bearing Pads

I Anti-Vibration Pads

J Composite Beams

K Quality Control (Source and Field)

L Survey (as built)

EXHIBIT 4    MASTER SPECIFICATION QUESTIONNAIRE

# 19

## Uniform Construction Index

With the advent of the *CSI Format for Construction Specifications* (Chapter 3), it became apparent to both CSI and AIA that this basic framework could be extended into a system for filing product data and to construction cost accounting as well as to a specifications outline.

A Joint Industry Conference Committee, in which many organizations from the United States and Canada participated, evolved the *Uniform System for Construction Specifications, Data Filing, and Cost Accounting, Title One Buildings,* which was published in October 1966. This document was a first attempt at correlating diverse elements in the construction industry so that a meaningful and useful agreed upon organized informational system could be achieved. It included the CSI Format which governs the organization of specifications; the AIA standard filing system which is concerned with the filing of manufacturers' literature; and *AGC Guide for Field Cost Accounting* which deals with cost analysis. The Conference on Uniform Indexing Systems brought these major organizations together and produced the Uniform System which provided a side-by-side arrangement of technical specification sections, manufacturers literature, and cost accounting so that there was a parallel arrangement of this information keyed to one another for easier classification and retrieval of this data. The Uniform System was primarily based on the organization of specifications with the filing of literature and the preparation of cost estimates tied to this arrangement.

In 1972 the Uniform System Joint Industry Conference revised the Uniform System and published the *Uniform Construction Index*. One of the major additions to the old Uniform System was the inclusion in the *Uniform Construction Index* of the Project Filing Format. This format is the contribution of the Canadian Building Construction Index Committee. The format is arranged so that it permits the filing of correspondence and information pertaining to a specific project, governing administration, design, bidding, and construction. However, the Project Filing Format is not keyed to the other three elements of the *Uniform Construction Index*. A separate matrix of broad scope headings is established that provides an overall format for the filing of project information.

The *Uniform Construction Index* creates a preferred terminology for section titles under the *CSI Format for Construction Specifications* (see Chapter 3) in a recommended sequence including a permanent number to all the section titles appearing in the *Uniform Construction Index*.

Of particular interest and significance to the specifier is the development within the specification outline of two categories of section titles: the broad scope title and the narrow scope title. This arrangement stems from the fact that since a specification section is essentially a unit of work, it can be very broad for one project and encompass many small items, or it can be very narrow in scope for another project and cover in detail a large volume of identical work. This permits the specifier the same flexibility he always enjoyed in the content of his technical section prior to the advent of the CSI Format and the *Uniform Construction Index*.

Some changes have occurred between the issuance of the Uniform System and the *Uniform Construction Index*. Some minor changes involve the names of the division headings as follows:

| *Former Name* | *Present Name* |
|---|---|
| Division 5 Metals; Structural and Miscellaneous | Metals |
| Division 6 Carpentry | Wood and Plastics |

Division 7  Moisture Protection          Thermal and Moisture
                                                        Protection
Division 8  Doors, Windows, and Glass   Doors and Windows

Other changes involve the rearrangement, addition, and deletion of some items. Obviously experience with the first Uniform System has dictated change which is reflected in the *Uniform Construction Index*.

A major change has occurred in Division 1, General Requirements (see Chapter 12), so much so that any specifier following the system has to rewrite this portion of his specification drastically. A side-by-side comparison is shown as follows:

| Uniform System (previous system) | Uniform Construction Index (present system) |
|---|---|
| Summary of Work | Summary of Work |
| Schedules and Reports | Alternatives |
| Samples and Shop Drawings | Project Meetings |
| Temporary Facilities | Submittals |
| Cleaning Up | Quality Control |
| Project Closeout | Temporary Facilities and Controls |
| Allowances | Material and Equipment |
| Alternates | Project Closeout |

The use of the long recognized term "alternate" has been replaced by the term "alternative," which may not be as readily acceptable as the old term. The placing of Samples and Shop Drawings under Submittals, and Allowances under Summary of Work are improvements, but locating Cleaning Up under Project Closeout is inappropriate. Project Closeout should be reserved solely for documentation attesting to receipt of information such as guarantees, bonds, record drawings, release of liens, maintenance manuals, certificates, and so on, so that the architect or engineer can sign a final certificate for payment.

The creation of a separate title "Material and Equipment" is sound since generalized information on transportation and handling, and storage and protection, can be specified here once with little need to repeat it in every technical section as has been the custom.

The new heading "Quality Control" similarly provides a convenient place to spell out the scope of the contractor's quality control program.

There are some discrepancies with respect to the Specifications Format Part 1, of the *Uniform Construction Index* and the CSI Format, Document MP-2A, dated September 1972. The *Uniform Construction Index* uses the term "Material and Equipment" as a section title under Division 1; whereas the CSI Format uses the term "Products." Other differing terms similarly appear in both documents, especially for narrow scope section titles. In some instances the order of the section titles is not consistent.

The *Uniform Construction Index* can be obtained from either of the following organizations:

American Institute of Architects
1785 Massachusetts Avenue
Washington, D.C. 20036

Construction Specifications Institute
1150 Seventeenth Street, N.W.
Washington, D.C. 20036

# 20

## *Physical Format and Style*

1 FORMAT AND ARRANGEMENT
   a. The format and arrangement of specifications are essential to a workable system. An architect carefully plans his layout of drawings. Plans, elevations, sections, details, and schedules are usually laid out on the drawings in an orderly fashion and the drawing numbers also follow in certain sequences.
   b. An architect should similarly give some thought to the appearance of his specifications and provide for some system in their physical makeup.

2 SPECIFICATION SECTIONS
   a. The section number and section title can be arranged, as illustrated for this chapter heading, or may appear on one line, as follows:
      *CHAPTER 20   PHYSICAL FORMAT AND STYLE*

3 PARAGRAPH HEADINGS
   a. In Chapter 5 the observation is made that a technical section in a book of specifications could be considered as analogous to a chapter in a book, and that the breakdown within the section, as in a book, consists of paragraphs and subparagraphs.
   b. Each major paragraph heading should be capitalized and underscored and be preceded by the section number, followed by its paragraph number as illustrated for this paragraph:
      20.3  PARAGRAPH HEADINGS.

**4 SUBPARAGRAPHS**

  a. <u>Subparagraph Headings</u> under the major paragraph, such as this one, are written in lowercase, continuously underscored with only the first letter of each word capitalized.

  b. <u>Other Paragraph Headings</u>: When the heading does not lend itself to a part of the sentence, end the heading with a colon and write the paragraph in accordance with the following illustration:

    1. The first subparagraph should be indented and numbered in accordance with this illustration.

      (a) A subparagraph would conform to this illustration.

        (1) A further subdivision of the paragraph would follow this illustration.

  c. It is recommended that specifications be written so that there is no necessity to go beyond the order of the subparagraph shown in Par. 4b.1(a)(1). Each paragraph, subparagraph, and sub-subparagraph should be numbered as illustrated to facilitate reference to the specific paragraph in question during correspondence, in addenda, or in discussion. By utilizing the term "paragraph" only, one is not burdened with other systems of nomenclature that refer to articles, headings, paragraphs, categories, and items.

**5 BLOCK STYLE**

  a. This style of typing and paragraphing is recommended for specifications since it is easy to read and permits quick recognition of information when it is sought. This chapter is written in the block style. See samples at the end of this chapter.

**6 PAGE NUMBERING**

  a. <u>Page Numbers</u> should be used for ready reference. Continuous numbering from cover to cover is a most difficult and a time-consuming task, and since last minute changes can upset the entire effort, it is recommended that the pages of each section be numbered independently of the others. It is suggested that the page numbers consist of the section number followed by numbers in sequence. The first page of Section 1A would therefore be 1A-1, and succeeding pages of that Section would be 1A-2, 1A-3, and so on.

  b. <u>Location of Page Numbers</u>: It is recommended that the page number be located at the bottom center of the page. Others may prefer to use the lower right-hand or upper right-hand corner, which is often used in books.

7  CLOSING NOTE
  a. Endings of each section should conclude with a symbol or mark that signifies the end of the section, such as

***

  or

**END**

## SECTION 05510
## METAL STAIRS
## PART 1   GENERAL

### 1.1 SCOPE

a. <u>Work Included</u>: Provide steel pan-type stairs as shown or specified, and in accordance with requirements of the contract documents.

b. <u>Work Of Other Sections</u>
   1. Cement fill for steel pan-type stairs.

### 1.2 SUBMITTALS

a. <u>Shop Drawings</u>: Show in detail the construction, gauges, jointing, methods of installation, fastening and supports, location and sizes of welds, anchoring, hangers, and all other pertinent data. Submit details drawn to scale at not less than $1/4$ in./ft. Show design, type of steel and load assumption, and note the seal of a licensed professional engineer registered in the state in which the project occurs.

b. <u>Samples</u>: Submit samples of materials or assemblies, as requested.

## PART 2   PRODUCTS

### 2.1 MATERIALS

a. <u>Structural Steel</u>: ASTM A36

b. <u>Steel Sheets</u>: ASTM A570 or A 611 grade as required to meet structural requirecents.

c. <u>Steel Pipe</u>: ASTM A53, Type E, Grade A, suitable for close coiling, black, unless otherwise shown as galvanized.

EXHIBIT B    SHOWING INDENTED FORM

## SECTION 05510
## METAL STAIRS
### PART 1    GENERAL

1.1 SCOPE

(a) Work Included: Provide steel pan-type stairs as shown or specified, and in accordance with requirements of the contract documents.

(b) Work Of Other Sections

1. Cement fill for steel pan-type stairs.

1.2 SUBMITTALS

(a) Shop Drawings: Show in detail the construction, gauges, jointing, methods of installation, fastening and supports, location and sizes of welds, anchoring, hangers, and all other pertinent data. Submit details drawn to scale at not less than ¼ in./ft. Show design, type of steel and load assumption, and note the seal of a licensed professional engineer registered in the state in which the project occurs.

(b) Samples: Submit samples of materials or assemblies, as requested.

### PART 2    PRODUCTS

2.1 MATERIALS

(a) Structural Steel: ASTM A36.

(b) Steel Sheets: ASTM A570 or A611 grade as required to meet structural requirements.

(c) Steel Pipe: ASTM A53, Type E, Grade A, suitable for close coiling, black, unless otherwise shown as galvanized.

EXHIBIT C    SHOWING FLUSH LEFT FORM

## SECTION 05510
## METAL STAIRS
## PART 1    GENERAL

1. SCOPE
a. Work of Other Sections
(1) Cement fill for steel pan-type stairs.
2. SUBMITTALS
a. Shop Drawings: Show in detail the construction, gauges, jointing, methods of installation, fastening and supports, location and sizes of welds, anchoring, hangers, and all other pertinent data. Submit design, type of steel, and load assumption, and note the seal of a licensed professional engineer registered in the state in which the project occurs.
b. Samples: Submit samples of materials or assemblies, as requested.

## PART 2    PRODUCTS

1. MATERIALS
a. Structural Steel: ASTM A36.
b. Steel Sheets: ASTM A570 or A611 grade as required to meet structural requirements.
c. Steel Pipe: ASTM A53, Type E, Grade A, suitable for close coiling, black, unless otherwise shown as galvanized.

# 21

## *Addenda*

### *Definition*

This last chapter is an appropriate place to discuss Addenda, since by their very nature, Addenda are intended as clarifications of previously issued instructions. The dictionary definition of an addendum is "a thing to be added; an addition." Each addendum is a document added to a previously prepared and issued set of contract documents during the bidding period, and it becomes a part of the contract documents as defined by *AIA General Conditions*, Article 1.1.1.

### *Purpose*

In Chapter 9, under the heading "Instructions to Bidders," paragraph 5 is concerned with the interpretation of documents. The primary purpose of an addendum is to clarify, in writing, questions raised by bidders as to the meaning of the drawings and specifications, or to discrepancies or omissions therein, prior to the receipt of bids. In addition, the addendum can be used as the instrument or vehicle by which additional information is made a part of the contract documents. This additional information can take any of the following forms:

1. Correct errors and omissions
2. Clarify ambiguities
3. Add to or reduce the scope of the work
4. Provide additional information that can affect the bid prices
5. Change the time and place for receipt of bids
6. Change the quality of the work
7. Issue additional names of qualified "or equal" products (see *Product Approval Standards* under Chapter 13, "Specifying Materials," and Chapter 12, "General Requirements"

## Language of Addenda

Clarification by addenda of discrepancies, ambiguities, errors, and omissions in the original contract documents should not be in the form of explanations. They should be precise instructions, using the language previously employed in the documents. For example, if the drawings show asphalt tile and the specification describes vinyl asbestos tile, one of these documents must be changed. If vinyl asbestos is actually required, the addendum should state the drawing on which the information is contained and instruct that the term "asphalt tile" be deleted and the term "vinyl asbestos tile" be inserted. If asphalt tile is actually required, the specification should be altered by the addendum to state that vinyl asbestos tile be deleted and asphalt tile be inserted, together with the appropriate reference to the proper material. In essence, the altered documents should read as original documents.

If there are conflicts on the drawing, in which one detail shows one arrangement and another detail shows another arrangement, delete the inappropriate detail by addendum rather than explain that it is intended that one detail governs over another. Similarly for conflicts within the specifications, delete the inappropriate material by addendum; do not explain that one is preferred or should govern over the other.

Where changes by addendum to a paragraph of the specifications may be misleading or ambiguous, delete the entire paragraph and rewrite it by addendum. The cardinal rule is to use specification language in the addenda by instructions to the contractor rather than by explanation.

## Precautions

Do not issue an involved addendum that requires considerable work on the part of the bidders unless there is still sufficient time before the bid due date. If time is scant for such a change, alter the bid due date.

Do not clarify a bidder's telephone inquiry verbally. Instruct him to reduce the inquiry to writing, and if a clarification is in order, answer it by an addendum to all bidders so that every bidder is informed through the right channels and there is no misinterpretation of verbal instructions.

A simple change two or three days prior to receipt of bids can be made by telegram; however, the content of the telegraphic addendum should then be transmitted in the form of a formal addendum for the record.

In private work, where changes are negotiated with a successful bidder before award of contract, the changes should be incorporated in the form of an addendum prior to the execution of the agreement or the contract form.

All addenda should be prepared, controlled, and issued by one individual, preferably the specification writer, who has intimate knowledge of all the contract documents and who serves as the clearing house for gathering and arranging all the bits of information. Do not permit consultants, that is, structural, mechanical, and site engineers to issue addenda since they may inadvertently assign wrong addenda numbers or issue instructions that conflict with other instructions contained elsewhere in the contract documents.

Itemize each instruction or change within the addendum by a number for future reference during construction and correspondence. (See sample addendum at the end of this chapter.)

## *Format of Addenda*

Each addendum should be arranged in an orderly sequence. Changes to the Project Manual (book of specifications) should follow the sequence of the table of contents, after the introductory statements, as follows:

### Introduction

1. Name of architect, engineer, or issuing agency
2. Project identification
3. Addendum number
4. Date of addendum
5. Opening remarks and instructions (see sample addendum)

### Sequence of Addendum Changes

1. Changes to prior addenda
2. Changes to table of contents
3. Changes to Invitation to Bid
4. Changes to Instructions to Bidders
5. Changes to Bid Form
   *Note*. Reissue separate pages of the Bid Form requiring changes
6. Changes to contract forms
7. Changes to General and Supplementary Conditions
8. Changes to specification sections—in sequence
9. Changes to drawings—in sequence

# JOHN JONES, ARCHITECT
## 123 MAIN STREET
## NEW YORK, NEW YORK

### LIBRARY BUILDING
### FIRST AVENUE and MAIN STREET
### NEW YORK, NEW YORK

### ADDENDUM NO. 2
### January 10, 1974

The original specifications and drawings, date January 1, 1974, for the project noted above are amended as noted in this Addendum No. 2.

Receipt of this Addendum shall be acknowledged by inserting its number and date in the space provided on the Bid Form.

This Addendum consists of _____ pages (and the attachments noted herein).

## ADDENDUM NO. 1

**ITEM NO.**

ADD 2-1     Item No. ADD 1-3. Add the following: "Mirrors shall be Model A as manufactured by XYZ Company."

## BID FORM

ADD 2-2     *Delete page BID-3 bound in the specifications dated January 1, 1974. The attached page BID-3 (REVISED, ADDENDUM NO. 2) shall be used by all Bidders.

## SPECIFICATIONS
### SECTION 4A, UNIT MASONRY

ADD 2-3     Page 4A-2, Par. 4A.3.b., Line 2, Change "$60/M cash allowance" to "$80/M cash allowance."

### SECTION 9A, RESILIENT FLOORING

ADD 2-4     Page 9A-5, Par. 9A.7.c.1., Delete last sentence.

## DRAWINGS
### DRAWING A-3

ADD 2-5     Room 302. Change flooring from "terrazzo" to "asphalt tile."

* Changes in Bid Form should be made by reissuing the specific page requiring a change. This is in keeping with the rule that the format of the Bid Form is best prepared by the architect.

# Index

**235**